Community Seed Bank Options

ISBN 978-1-946263-34-6

Edited by Patrick Trail, Dr. Timothy Motis, Jonathan Ribich, Dr. Abram Bicksler, and Stacy Swartz

Cover and interior design by Andre Fonseca
Artwork by Arielle Price

ECHO
Asia

ECHO Asia Impact Center
159 Moo 2
San Pa Pao, San Sai,
Chiang Mai, 50210 Thailand
Email: echoasia@echonet.org

ECHO is a non-profit Christian organization. For further resources, including networking with other agricultural and community development practitioners, please visit our website: www.ECHOcommunity.org. ECHO's general information website can be found at: www.echonet.org.

Contents

Foreword

From the outset of ECHO's founding in 1981, the provision of seed, scion wood, and other germplasm was decided to be a major area of need and opportunity for ministry. The idea of growing plants, saving seeds, and placing those seeds into the hands of those who needed them most took root and grew into a major ministry thrust of ECHO, starting in the United States and later expanding across the world. In 2009, with the launch of the first ECHO center outside of the USA, based in Thailand, the opportunity to localize seed banking approaches to the needs and contextual resources of the region was seized upon.

From there, ECHO's seed banking became a foundational component of the organization's DNA, with every Regional Impact Center establishing its seed banks. While the idea was to provide appropriate seeds and germplasm to our network partners worldwide, especially those of Neglected & Underutilized Species (NUS), it grew into so much more. ECHO's seed banking work became a way to contextually research, experiment, learn, share, equip, and serve those on the front lines of climate adaptation and mitigation, as well as agrobiodiversity and food culture preservation worldwide.

As the world has changed, ECHO's paradigm of seed banking has also changed: from operating seed banks that share small trial packets of seeds with network partners (which is still ongoing) to learning, experimenting, and sharing contextually relevant, low-cost methodologies of seed production, cleaning, drying, preservation, regeneration, and commercialization in seed systems. ECHO's paradigm has transformed to develop a global seed banking network of partners so that their capacities can be built and they themselves can determine and implement community seed banks based on local cultural and ecological needs and assets.

In your hands, you hold the culmination of 40+ years of learning, sharing, and research- dedicated to you- the seed bank managers of the world. We hope and pray that this work will be extremely practical in your fight for food sovereignty, local knowledge, biodiversity preservation, and food security through better seed preservation and utilization.

I once heard it said, "seeds are life," and they preserve food culture, biodiversity, local identity, and the very livelihoods of farmers and those who eat food. Local seed banks are critical in the fight against hunger and the preservation of life, making ECHO's work in global seed banking even more important. Thank you for joining us in this fight and championing the cause of a global seed bank network!

May this book be a blessing to you and your communities,

Abram J. Bicksler, Ph.D.

President and CEO, ECHO, Inc.

Preface

By Patrick Trail, ECHO Asia Regional Impact Center Director

Prior to my position as a research coordinator working with ECHO Asia's Small Farm Resource Center & Seed Bank, I had this notion that success in research meant discovering one big idea that could change the world. I have since evolved in my thinking to realize that good research is a culmination of many small research questions answered over time, with the aim of building up a better understanding of a subject matter from multiple angles and individual experiments. This, in a nutshell, is the premise of this publication. It is the compilation of many individual research experiments, papers, ideas, technologies, and experiences offered over many years, from many individuals, all working to further best practices in the realms of community seed banking, seed saving, and genetic preservation.

The idea for this publication came with the realization that ECHO's large collection of seed related work needed a home, one place to integrate it all. Building on ECHO's prior collections of agricultural and development best practices in *Amaranth to Zai Holes* (Meitzner and Price, 1996) and *Agricultural Options for Small-Scale Farmers* (Motis and Berkelaar, 2012), this publication follows a similar format but focused on one of ECHO's greatest strengths as an organization... seeds. With dozens of *ECHO Development Notes*, *ECHO Asia Notes*, peer-reviewed journal articles, and how-to-manuals devoted to small-scale seed banking practices, it was time to bring them together to form a comprehensive collection, thus the idea for '*Community Seed Bank Options*' was born.

I should note that this publication is reflective of ECHO's core network model, whereby new techniques, technologies, and best practices have been collected from a variety of individual practitioners and sources from all over the world; people that have come up with innovative solutions to complicated challenges willing to share it all with open hands. While ECHO has worked hard to collect and verify these best practices over many years, it should be made clear that we do not take credit for each individual innovation. We have remained committed to giving credit where credit is due and hope that this is reflected in each appropriate instance.

Intended Use(s) of This Resource

It is important that the reader understands that this publication has been carefully designed to accomplish a few different objectives, aimed primarily at current or prospective community seed bank managers. It is the authors' hope that this publication can be used as a practical tool in the following regards:

1). **As a Reference Guide:** Whether you are a current manager of a community seed bank or are hopeful of operating your own in the future, this book may serve as a technical reference guide. Readers will note that reference tables and appendices have been included to summarize detailed information such as acceptable seed germination rates for specific crop species, recommended planting distances for maintaining seed purity, and other such information that need not be committed to memory but might be referenced from time to time as needed.

2). **As a Collection of Practical Options:** While many scientific publications on the subject already exist, this book is aimed as a practitioner's guide for those working in resource-constrained contexts. The techniques, technologies, and best practices presented in the following pages have been carefully selected for their practical usefulness, as 'options' for those seeking to affordably operate their own seed saving entities.

3). **As a Training Curriculum:** This book has also been arranged in a format that can be used as a curriculum 'checklist' for community seed bank manager students and interns. This book is laid out in a way that takes a reader 'from A to Z' through the individual steps needed to operate one's own seed bank. From selecting crop varieties, to harvesting, drying, cleaning, storing, inventorying, and sharing, a student of the subject may use this as a guide for covering each necessary step.

With that said, I trust you will benefit from this practical toolkit and hope that it serves you well in your endeavors to save and share seeds with those that need them most. ECHO welcomes you into its growing network of community seed banks and looks forward to learning from your own innovations as you discover them. May this resource guide you as you serve others with access to seeds, a diversity of varieties, preservation of what is being lost, and improved food security for those you encounter.

Seed Learning Lessons Checklist

Introduction to Community Seed Banks

Chapter 1. The Global Seed Situation

- — Defining the Community Seed Bank
- — Smallholder Access to Improved Seed
- — A Bridge Between Formal and Informal Seed Systems
- — Why Community Seed Banks?

Chapter 2. Types of Community Seed Banks

- — In Terms of Function
- — In Terms of Operational Model

Part I. Seed Multiplication

Chapter 3. Botany Basics

- — Classifying Plants in the Seed Bank
- — Basic Plant Reproduction
- — Genetics

Chapter 4. Seed Biology Basics

- — Seed Biology
- — Orthodox Seed vs. Recalcitrant Seed
- — Factors Affecting Seed Storage

Chapter 5. Seasonality and Field Planning

- — Crop life cycle
- — Annuals
- — Biennials
- — Perennials
- — Day Length Sensitivity
- — Field Planning

Chapter 6. Plant Pollination

- — Pollination and Fertilization
- — Self-Pollination and Cross-Pollination
- — Open-Pollinated (OP), Hybrid, or GMO?

MeTL
GROUP
The People's Brand
www.metl.net
Manufactured by:
EAST COAST OILS AND FATS LIMITED
P.O.Box 50054, Zamcargo Street, Kurasini
Dar es Salaam, Tanzania
Tel: +255 22 2850860, 2851175, Fax:
E-mail: sales@metl.net

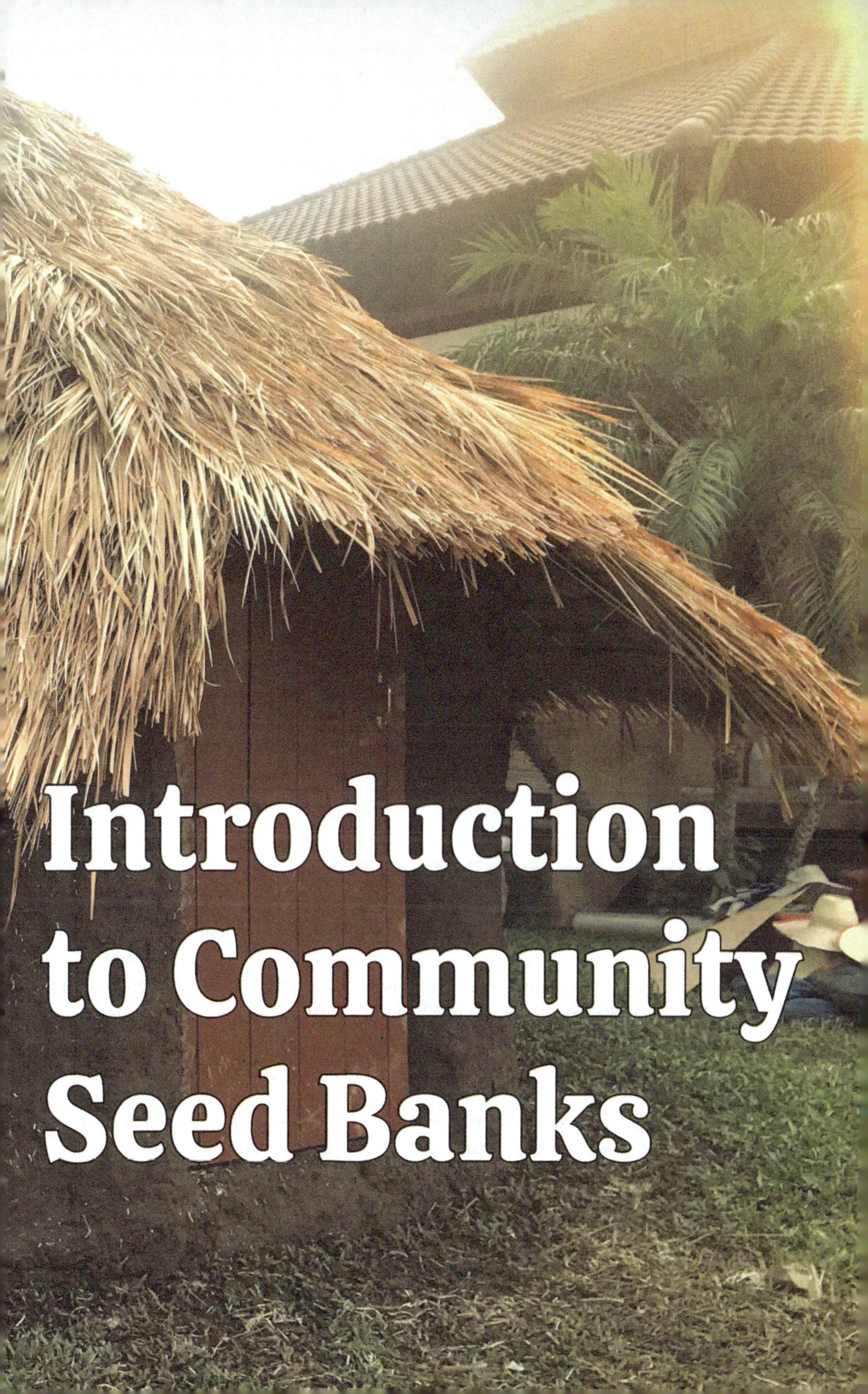
Introduction to Community Seed Banks

Chapter 1.
The Global Seed Situation

Seeds form the foundation of agricultural production, allowing growers to reproduce the diverse array of crops that feed our world. Although seed saving has been an integral part of agricultural production systems from the beginning, it faces new challenges that deserve attention. These challenges stem from a combination of environmental, social, and political constraints, in a rapidly changing global agricultural context. Farms are getting bigger, there are fewer farmers, and multinational companies are providing what was once produced locally. Traditional knowledge is being replaced with regimented agricultural approaches, including the adoption of hybrid seeds that farmers cannot save and multiply without a loss in crop performance over time. Seeds of improved genetics have become commercially available to many, but local genetics are being lost and commercially available seeds are often too expensive for those that need them most.

Nonetheless, smallholder farmers still have the opportunity to produce, save, and share seeds that are affordable and accessible to all. This book lays out practical tools for practitioners and field personnel serving smallholder farming communities. It presents a combination of background and scientific theory, coupled with tested techniques and technologies designed for establishing and operating small-scale seed banks in tropical climates with limited resources.scale seed banks in tropical climates with limited financial resources.

Defining the Community Seed Bank

First, we will define a few key terms to distinguish the many overlapping and sometimes confusing subjects at hand. What is the difference between a "seed bank" and a "gene bank"? Are they not the same? Are "seed saving" and "seed banking" interchangeable terms? Should these topics be approached differently? The short list of definitions below answers these questions. The following terms will be used repeatedly throughout this publication and will help you better understand the aims of this book:

> **Gene Bank:** A space or place where genetic material of potential importance can be stored for the future, typically for use by plant or livestock breeders and scientists. Genetic materials may include seeds, plant tissue cultures, germplasm, or even livestock breeding material.

Seed Bank: A space or place where seeds or plant germplasm of potential importance can be stored for the future. A seed bank is a type of gene bank that serves various stakeholders, including gardeners, farmers, crop breeders, and scientists. Seed banks may exist to preserve small quantities of many different species or larger amounts of only a few species.

Seed Saving: The practice of saving seeds for future reproduction or multiplication; traditionally practiced by individuals on a year-to-year basis, storing seeds between harvest and planting the following growing season. However, seed saving can also take place over a number of years, storing seeds for more extended periods.

Seed Banking: The practice of establishing, organizing, and operating institutions designed to save and store seeds. Seed banking involves the management of seed inventories, finances, and maintenance of seed storage infrastructure. Seed banking often operates beyond the individual level, typically organized and managed by multiple stakeholders.

Seed Company: A business focused on the reproduction and sale of seeds for profit. Seed companies typically focus on a limited number of crops and produce larger amounts for sale and distribution to buyers.

Community Seed Bank (CSB): A type of seed bank that exists to serve an individual community or small collection of communities. Community seed banks focus on providing access to higher quality seeds in areas underserved or unreached by formal seed systems, i.e., in remote or rural areas. Community seed banks are often owned or operated by members of the community within which they are located.

Community Seed bank (CSB)

Smallholder Access to Improved Seed

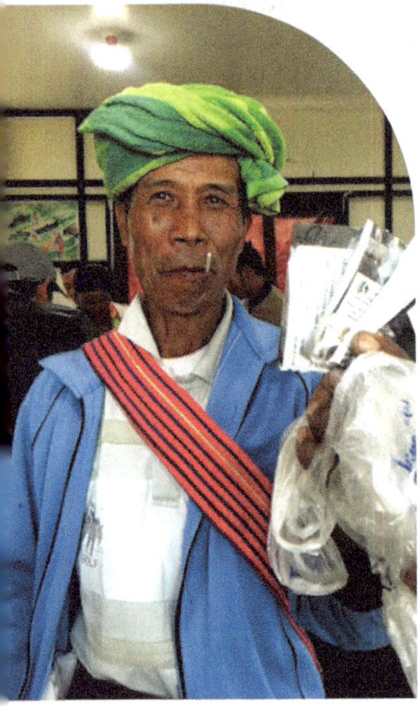

A Pa-O man holds up seeds he received during a seed exchange event in Myanmar.

Traditional storage of seeds over the fire, northern Thailand.

As previously mentioned, seeds are the foundation of agriculture, and their role in agricultural development cannot be understated. In a global agricultural landscape where 500 million smallholder farmers still operate, significant opportunities remain for improved access and use of locally available seeds. Access to improved seeds offers the smallholder farmer options for increasing the diversity of crops planted, flexibility in the timing of plantings, and overall yield improvement.

Yet, according to the 2019 Access to Seeds Index Report (World Benchmarking Alliance, 2019), only 47 million of the world's 500 million farmers had access to one of the world's 13 largest seed companies in 2017, less than 10% of smallholder farmers globally. So, where do smallholder farmers typically source their seeds? Are smallholder farmers buying their seed from markets and smaller seed companies, or are they saving seed themselves? These are critical questions, and together they form the foundational motivation for establishing small-scale, localized seed banks, from here on out referred to as Community Seed Banks or CSB's.

A Bridge Between Formal and Informal Seed Systems

Before considering the establishment of any seed bank or seed saving institution, it is essential to first understand the current seed systems of smallholder farmers in the target communities. Smallholder farmers tend to source their seeds from diverse avenues, generally belonging to two overarching systems: the formal seed system and informal seed system (Almekinders and Louwaars, 2002). As illustrated below, these two systems often operate independently from each other, and play significantly different roles depending on seed type, access, and cost to the farmer. Both have their place in the overall agricultural production systems and can benefit farmers differently.

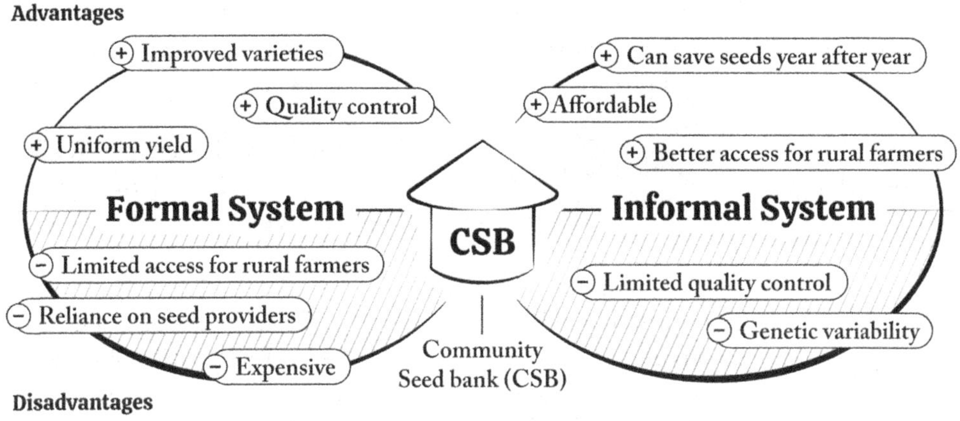

The local and formal systems of farmers' seed supply: two parallel functioning systems with relatively little interaction

The formal seed system is the more recognized and easily defined system. Seeds stemming from this type of system are typically produced by government or corporate institutions, resulting in "improved" or "modern" crop varieties. These often include hybrid seed varieties, most commonly of grain crops or high-value vegetable species. Formal seed systems typically involve international and country-level gene banks and crop breeding programs produced through university and research institutions. These systems place a premium on quality control, genetic purity, and specific crop characteristics, such as disease resistance, drought tolerance, and improved yield potential.

The informal seed systems, however, are not as easily defined but play a critical role in the agricultural production systems of smallholder farmers. In addition to traditional farmer seed saving practices, seed sourcing in these informal seed systems also includes exchange with neighbors and purchase from local seed vendors.

Seeds sourced through formal and informal seed systems are compared in the following table.

	Formal Seed Systems	Informal Seed Systems
Source	Government seed programs; national & international gene banks	Farmer seed saving; farmer exchange; local seed vendors
Percentage of Total Seeds	~10%	~90%
Farmer Access	Officially recognized seed dealers or government seed programs	Local seed vendors; neighbors; community networks
Type of Crop	Hybrid grain crops or high-value vegetables	Open-pollinated; heirloom; neglected & underutilized species
Benefits	Varietal uniformity; improved yield; targeted disease resistance; targeted crop characteristics; high quality control	Affordable; Seeds can be saved year-to-year; locally adapted; locally available; genetic diversity
Limitations	Expensive; limited genetic diversity; seeds cannot be saved year-to-year	Low quality control; uncertain market/ availability

In this book, we will focus on CSBs, which offer great potential to bridge the gaps between the formal and informal seed systems. With the ability to provide higher quality seeds and improved genetics to rural communities, CSBs offer the potential for much-needed services that may not exist. Though CSBs resemble formal seed systems by keeping record of seed varieties and maintaining quality control, they operate informally to share, trade, and disseminate seeds with those in underserved communities. This balance makes CSBs increasingly attractive to rural communities and partially explains their increased prevalence worldwide.

Why Community Seed Banks?

There are many benefits to establishing and operating CSBs and many incentives for incorporating them into rural development strategies. CSBs have the potential to combat the loss of crop genetic material and therefore slow the loss of crop biodiversity. Ongoing biodiversity loss is caused by many factors, including a global decline in the number of farmers and their traditional knowledge; increased planting of hybrid crops and subsequent reliance on seeds companies; and market preferences that favor transportable or marketable crop species. Of the world's nearly four hundred thousand identified plant species, over 50% of the world's calories come from just three crops – rice, wheat, and maize (Vernooy et al., 2020). For this reason, it is important to promote strategies that protect and preserve local crop biodiversity, especially in places where once diverse diets are becoming increasingly homogenized.

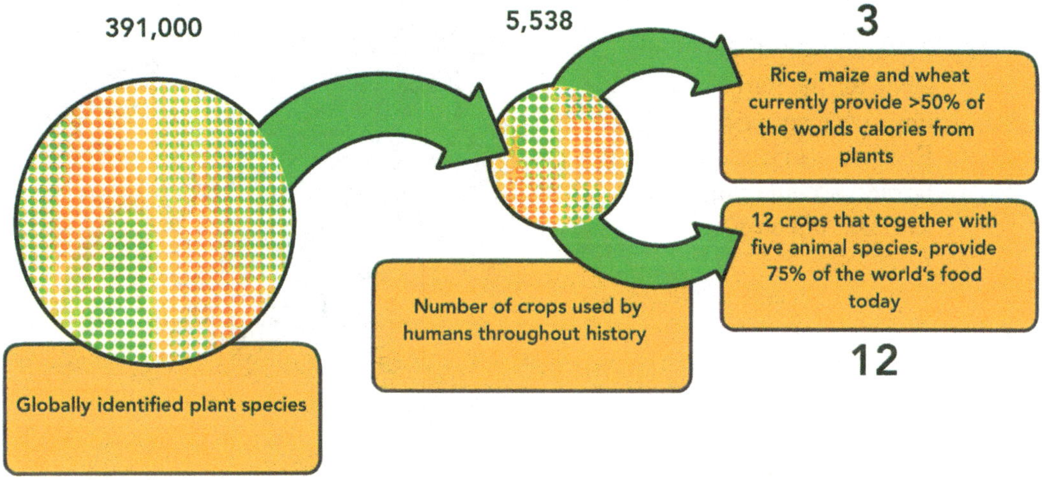

The State of the World's Plants (RBG Kew, 2016; FAO, 1999). Image adapted by Arielle Price

Additionally, CSBs provide benefits in the event of a major disaster. It is common for natural disasters to destroy farmers' fields and crop yields. These events include floods, typhoons, landslides, drought, civil conflicts, pandemics, strikes, and other disruptive forces. In the event of such a disaster, smallholder farmers must have immediate access to seeds for replanting and rebounding from calamity. CSBs may offer locally accessible seed when roads, infrastructure, and supply chains are down. This provides a unique resiliency mechanism for rural communities that may experience delays in receiving aid and relief.

Chapter 2. Types of Community Seed Banks

In Terms of Function

Seed banks, like seeds, are diverse. They differ in size, scale, and purpose. Categorizing them, however, is useful in planning community seed banking efforts. This can be done based on use, by sorting them into one of three identified overall seed bank functions:

1. Assistance: Exist to provide seed to farmers in need, either by sale or free distribution.

2. Production: Exist to offer seeds for the purpose of improving crop genetics. Recipients may include farmers seeking improved varieties and crop breeders seeking germplasm for improving seed varieties.

3. Preservation: Exist to maintain and protect seed germplasm. Seeds are not typically multiplied for sale or distribution, but for safeguarding against loss of genetic material in the event of calamity.

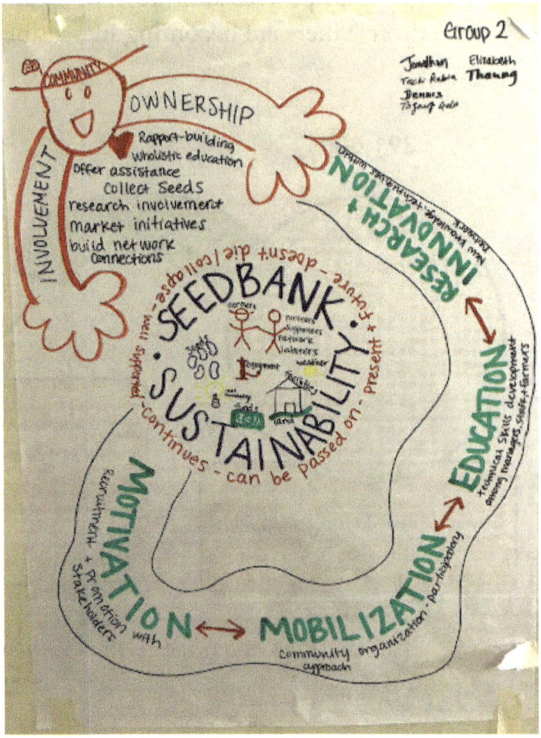

Results of a participatory learning activity at the first Regional Seed Bank Managers' Forum in Chiang Mai, Thailand, 2019.

CSBs may exist to offer a combination of these benefits to end users, but it is good practice to identify a primary purpose before the establishment of any operation (Pellegrini and Balatti, 2016).

In Terms of Operational Model

Institutional Model

Community Seed Banks of this model have traditionally been established through the support of non-governmental organizations (NGOs), with the aim of empowering local communities to preserve and share their seeds. The seed banks are located within the communities they serve but supported and funded through external sources. Increasingly, local grassroots organizations are supporting the establishment of institutional seed banks. Seeds are distributed to target communities for free, for sale, or for exchange, but are largely subsidized by the host organization.

Depending on resources, institutional seed banks can range from sophisticated cold storage facilities to low-cost, off-grid seed storage rooms.

Cooperative Model

Unlike institutional seed banks, cooperative seed banks are member-driven, whereby beneficiaries provide support through membership fees or commitments. Members can include individuals, or groups of individuals. One example is the Kahelu Seed Bank in Pathein, Myanmar. It is supported through a large association of churches, each contributing a small amount for the ongoing support of the seed bank. Similar arrangements exist with various farmer cooperatives that have placed value on seed preservation and access by its members.

Seed Libraries

A **seed library** is a model of seed bank that allows growers to obtain seeds for free, or through a membership, with the understanding that they will return some seeds from the resulting crop at the end of the season. In a sense, users are borrowing seeds. Seed libraries may operate in a manner that is open to the public or reserved for growers who have a membership with the seed bank. They typically serve the local population within a community and focus on open-pollinated, genetically diverse seeds that are suitable for growing in the region (Global Seed Savers, n.d.).

The Lisu Seminary Community Seed Bank, Pyin Oo Lwin, Myanmar.

Part I. Seed Multiplication

Chapter 3. Botany Basics

Classifying Plants in the Seed Bank

Here we distinguish between common names and botanical names. The common name of a plant refers to what it is commonly called locally. Plants have various common names in different languages, dialects, locations, or even for different uses (e.g. green vs. ripe bananas). While there can be many common names for a plant, a plant will only have one botanical name. Botanical names are also called scientific names.

Scientific nomenclature gives a foundation of classification that remains constant and is therefore used in seed bank management. Using scientific nomenclature also helps us make management decisions based on what we know about how different lines of plants are related to each other. All terrestrial, multi-cellular plants belong to the same overarching **Kingdom** of vascular plants. From there, plants that form seeds are in the **Division/Phylum** of flowering plants. Flowering plants are divided further into major Classes (dicotyledons, monocotyledons, etc.). The **Order** classification can be very helpful with plant ID because plants in the same order often have similar plant structures such as leaf patterns or inflorescences (flowering structures).

In seed banking, you will mainly utilize **Family**, **Genus**, and **Species** classifications.

Classifying using taxonomy

Kingdom	*King*	Plantae
Division	*David*	Magnoliophyta
Class	*Came*	Magnoliopsida
Order	*Over*	Cucurbitales
Family	*From*	Cucurbitaceae
Genus	*Germany*	Cucumis
Species	*Swimming*	Cucumis sativus

Plant classification tool based on highest-(kingdom) to lowest-order (species) taxonomic levels.

Family

Plants of the same family, such as Alliums in the Amaryllidaceae family, have similar requirements for management in the field, can typically be harvested in similar ways, and have similar sets of post-harvest processing and storage techniques. As you make seasonal plans to grow out seed for multiplication, rotate through plant families because plants in the same family will tend to have similar nutrient requirements, disease susceptibility, and overall growth habit.

Genus

A genus is a group of similar species and can have thousands of species/subspecies, like the *Vigna* genus. Plants of the same genus cannot cross with one another, with a few rare exceptions. These exceptions lead to intergeneric hybrids which are not genetically stable and require regeneration of hybrid seed for each year's planting.

Species

Plants of the same species have the genetic potential to reproduce or cross with one another. How plants reproduce (self-fertilization vs. cross-pollination) will dictate to what extent crossing actually happens. A plant species is always represented with an italicized Latin binomial in which the genus is capitalized, and the species is lowercased as shown below. Plants of the same species have more similarities than differences in their genetic makeup.

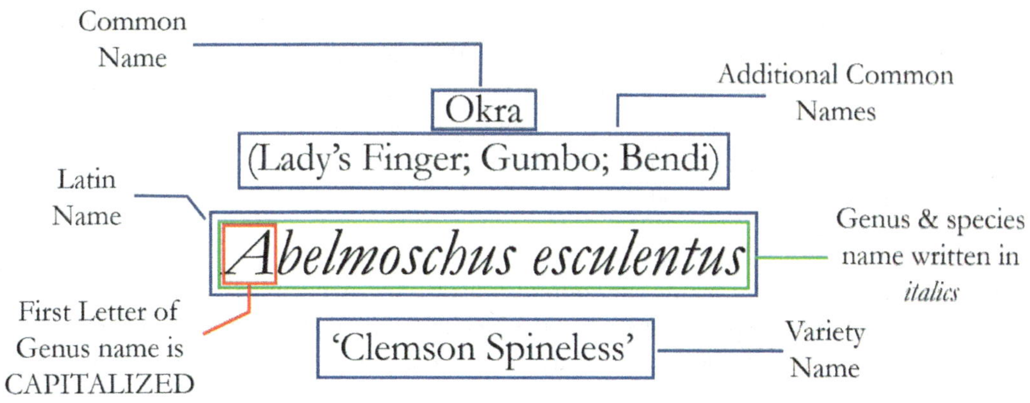

Elements of plant naming as illustrated by okra (*Abelmoschus esculentus*).

There are subspecies of species such as the diverse subspecies of *Brassica oleracea* (e.g. *Brassica oleracea* var. capitata is cabbage while *Brassica oleracea* var. botrytis denotes cauliflower).

The terms varieties, landraces, or cultivars are lines created from human selection or intervention and while not a part of technical nomenclature, help us define plant genetic developments and are equally important.

Basic Plant Reproduction

Vegetative versus seed

Plants reproduce asexually and/or sexually. Asexual reproduction involves the use of vegetative plant parts to propagate new plants. Examples of vegetative plant parts include above-ground stems (e.g. chaya), stems that run horizontally above (stolons; e.g. strawberry) or below (rhizomes; e.g. bamboo) the soil surface, and root structures such as tubers (e.g. yam) or modified underground stems called tuberous roots (e.g. sweet potato). Vegetative propagation methods include the taking and rooting of stem cuttings, root division, and grafting.

Seed bank personnel may choose to offer vegetative material, such as cuttings and tubers, to clients. Seed banks, however, deal primarily with plants that reproduce sexually. Sexual reproduction of plants involves the pollination of flowers to produce seeds. Most agricultural crops flower and set seed that is then dispersed by wind, water, air, or animals (including humans).

An understanding of pollination informs seed multiplication practices and is based on knowledge of flower parts.

Flower Structure and Function

The reproductive structure of an angiosperm (flowering plants) contains one or more flowers. Flowers may contain all parts or only some of the parts depending on the species. The **ovule** is located in the center of most flowers, at the base. Pollen reaches the ovule after being transferred from **stamen** (male flower structure) to **stigma** (female flower structure) and traveling down the **style**. Each ovary within the ovule produces a seed. A flower is said to be female if it only has female structures. Similarly, flowers with only male parts are said to be male. Flowers with either female or male structures are called **imperfect** flowers, while those with both male and female structures are referred to as **perfect**.

Some plants, such as pumpkins, are **monoecious**, having male and female flowers on the same plant. Other species, such as gac (*Momordica cochinchinensis*), are **dioecious**, having male and female flowers on different plants. This is important to know for seed production so that you can calculate the number of plants needed for successful pollination and genetic diversity. **Petals** can vary in color based on variety and can attract pollinators to the flower. The **sepals** (collectively called a calyx) protect flower buds, can also attract pollinators, and are sometimes edible (e.g. roselle).

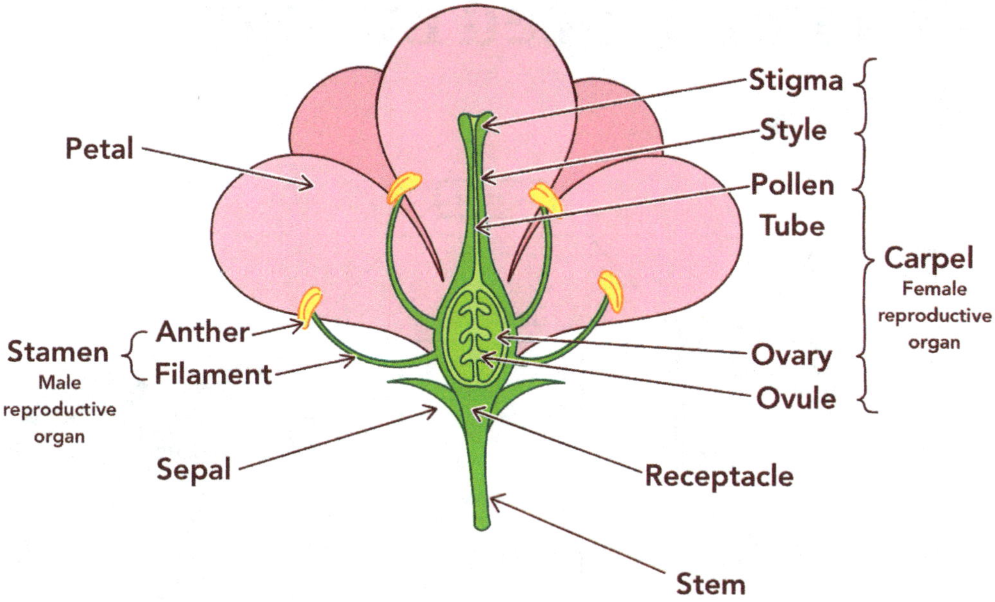

Flower structure diagram. Image source: Arielle Price

Chapter 6 contains more information about pollination and its implications for multiplying seeds.

Genetics

Plant traits are passed from one generation (parent) to the next (progeny). Traits are characteristics of a plant, such as flower color, that are controlled by **genes**. Each genetic trait may have complete, incomplete, or intermediate dominance and so may be variable in the likelihood that it is passed from one generation to another.

Vegetative propagation results in plants that are genetically the same as the parent plant. Seed propagation, on the other hand, results in offspring with a combination of genes from parent plants. Thus, plants grown from seed may not have the exact same traits as their parents.

Chapter 4. Seed Biology Basics

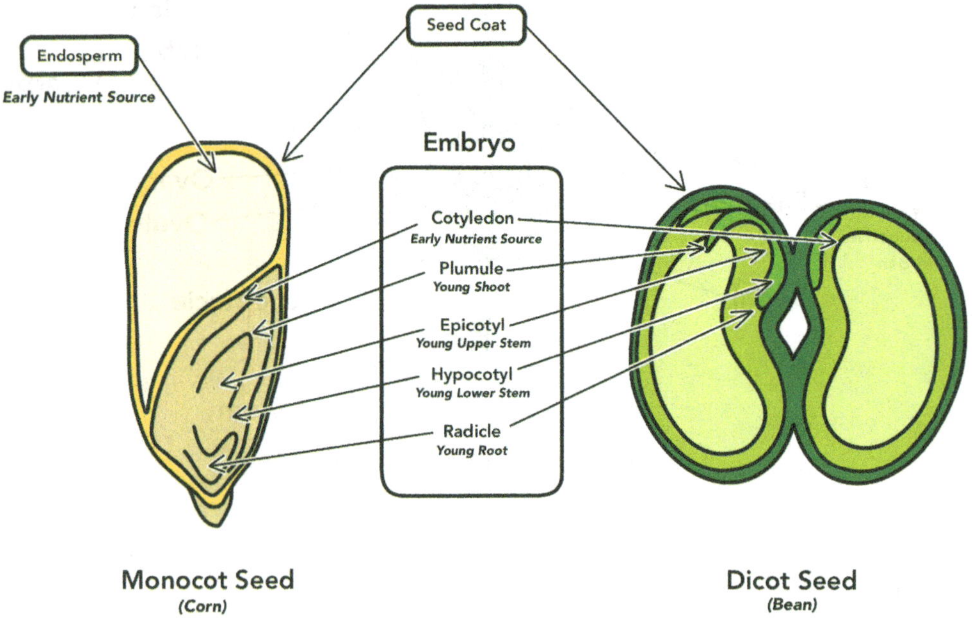

Monocot and dicot seed anatomy. Image source: Arielle Price

Seed Biology

Parts of a Seed

A seed consists of a living plant **embryo** surrounded by a protective **seed coat** and supplied with food reserves (the **endosperm**) to keep it alive and vigorous (Figure above). When a seed is fully mature, it has had enough time to develop a strong seed coat and build up sufficient food reserves. The plant embryo in a seed is the product of the fertilization of an ovule by pollen. When handling seeds, testing seed germination, or harvesting mature fruit be careful not to damage the areas of the seed containing the embryo. For example, when nicking the seed coat of beans during germination testing, always nick seed coat on the outside, rounded part of the seed where all you could damage would be the cotyledon (not the embryo).

Comparison of monocot and dicot structural differences. Image source: Arielle Price

Seed Respiration

Remember that seeds are alive. Seeds intake oxygen and respire carbon dioxide. As a seed consumes oxygen in a storage container, it uses some of its stored energy (glucose). Long-term storage practices such as vacuum sealing and cold storage slow respiration but do not stop it completely.

Glucose + oxygen = carbon dioxide + water

Tropisms

Auxin is a plant hormone that stimulates growth through cellular division. After a plant is seeded, first the seed **radical** (root) and then the seed **plumule** (shoot) emerge. A tropism is plant growth toward or away from a particular stimulus, usually through the production of auxin. Geotropism is plant growth in response to gravity. Phototropism is plant growth in response to light. Regardless of how you orient a seed in planting media, the radical will grow downwards while the plumule will grow upwards using one or both of these plant tropisms. When planting, as much as possible orient the seed so that it doesn't have to waste energy reserves in forming extra root or shoot tissue to grow the correct orientation.

Seed Dormancy

Seeds naturally enter a dormancy period after reaching maturity. This dormancy is broken when conditions for germination are met. The type of seed influences the length of time and conditions under which it can be stored, as well as seed loss.

Orthodox Seed vs. Recalcitrant Seed

Orthodox seeds slowly lose viability over time, should be kept at a low moisture content (10% or less), can be stored under freezing conditions, and tend to be small in size (e.g. grains, most vegetables). In contrast, **recalcitrant seeds** quickly lose viability, need to be kept moist, do not tolerate desiccation or freezing, and are usually large in size (e.g. avocado, jackfruit, mango). Some seeds are in between orthodox and recalcitrant in terms of storage life and tolerance to freezing or drying (e.g. citrus, coffee). Most of the content that follows relates to orthodox seed.

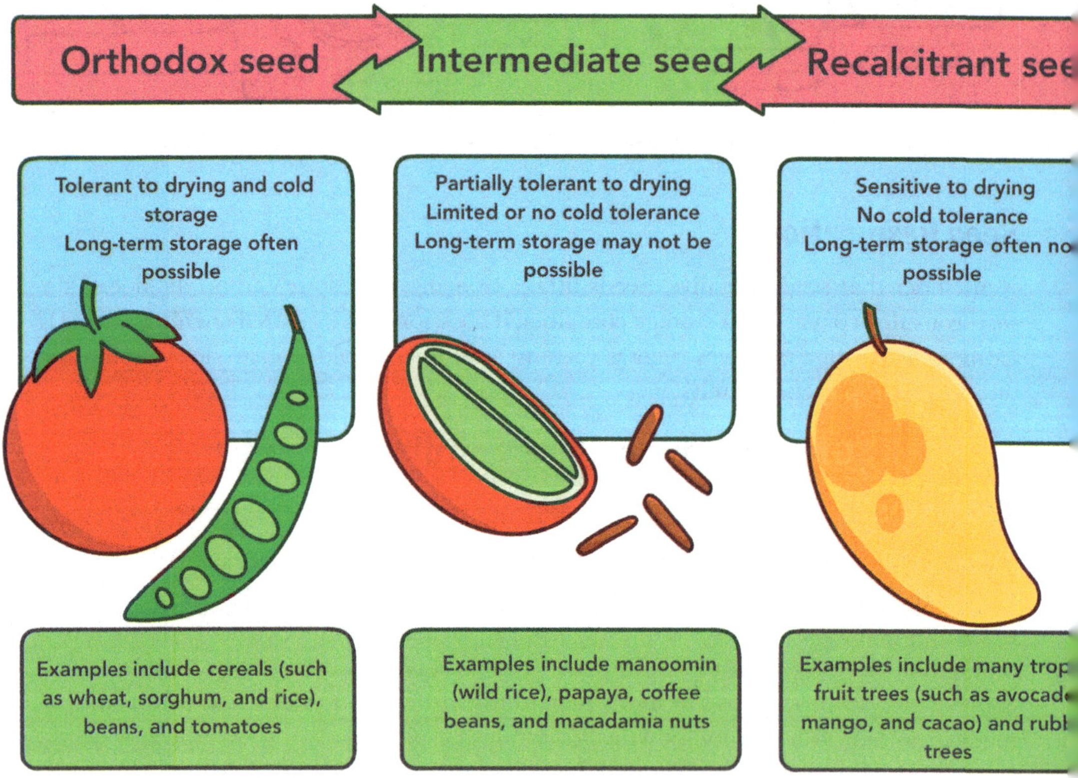

Comparison of orthodox, intermediate, and recalcitrant seeds (https://wildricebreedingandgenetics.umn.edu/seed-storage). Image adapted by Arielle Price

Factors Affecting Seed Storage

Seeds without innate dormancy germinate after exposure to adequate temperature, water, and oxygen, while seeds with innate dormancy require an additional stimulus (Frankland and Taylorson, 1983). When storing seeds, we want to minimize seed exposure to stimuli that initiate germination.

Moisture

Seeds need moisture to germinate but they should be kept dry for maximum storage life. Doing so slows down life processes, reducing the rate of seed deterioration and the degradation of food reserves that seeds need to grow when planted. Dry storage conditions also prevent the growth of mold and disease-causing organisms that thrive in humid environments.

Temperature

Temperature affects the speed of life processes. Seed germination and mold growth happen faster at high than low temperatures. Temperature also affects the amount of water vapor in the air, with more water vapor in warm than cool air. When warm air encounters a cold surface some of the water vapor it contains is released in liquid form as condensation. Moisture caused by condensation reduces storage life, which is why it is best to avoid moving seeds back and forth between cool and warm environments. Aim for low, stable temperatures when storing seeds.

Oxygen

Unlike moist seeds, dry seeds require little if any oxygen to remain viable. Minimizing oxygen in seed storage spaces extends seed storage life by slowing life processes and the accumulation of unstable molecules that damage seeds.

Light

Light is one potential stimulus that some crops require for germination to begin. Crops that require light to germinate include lettuce, celery, dill, and onion (Steil and Jauron, 2023). Aside from the fact that germination of some small seeds is stimulated by light, storing seeds in a dark place or container minimizes temperature gradients (e.g. the side of container facing a lighted window may experience higher temperatures than the opposite side).

Chapter 5. Seasonality and Field Planning

Crop Life Cycle

Angiosperms (flowering plants) can be **annuals**, **biennials**, or **perennials**. These terms indicate the length of time required from seeding to harvest of mature fruit and seeds. Knowing crop life cycles is important for planning as some plants will require longer growing periods, even multiple years, to produce seed for saving or distribution.

	Annual Plants	Biennial Plants	Perennial Plants
Complete a life cycle in...	One growing season	Two years	More than two years (continually produce)
Examples	Lettuce, tomatoes, pumpkins	Carrots, cabbage, onions	Mango, jackfruit, chaya
Growth and development	Germinate, develop, fruit, and die all in the same year	Germiante and grow leaves and stems in the first year. Flower and fruit in the second year.	Continue to produce leaves, flowers, and or/ fruit several years.

Annuals

Crop	Species	Family	Life Cycle
adzuki bean	*Vigna angularis*	Fabaceae	annual
amaranth	*Amaranthus* spp.	Amaranthaceae	annual
Armenian cucumber	*Cucumis melo*	Cucurbitaceae	annual
arugula (rocket)	*Eruca sativa*	Brassicaceae	annual
barley	*Hordeum vulgare*	Poaceae	annual

Crop	Species	Family	Life Cycle
bean (common bean)	*Phaseolus vulgaris*	Fabaceae	annual
broccoli raab	*Brassica rapa*	Brassicaceae	annual/biennial
buckwheat	*Fagopyrum esculentum*	Polygonaceae	annual
celtuce	*Lactuca sativa*	Asteraceae	annual
chickpea (garbanzo bean)	*Cicer arietinum*	Fabaceae	annual
Chinese cabbage	*Brassica rapa*	Brassicaceae	annual/biennial
cilantro	*Coriandrum sativum*	Apiaceae	annual
citron melon	*Citrullus lanatus*	Cucurbitaceae	annual
coriander	*Coriandrum sativum*	Apiaceae	annual
corn (maize)	*Zea mays*	Poaceae	annual
cowpea (Southern pea, yardlong bean, black eyed peas)	*Vigna unguiculata*	Fabaceae	annual
cucumber	*Cucumis sativus*	Cucurbitaceae	annual
currant tomato	*Solanum pimpinellifolium*	Solanaceae	annual
dill	*Anethum graveolens*	Apiaceae	annual
eggplant	*Solanum melongena*	Solanaceae	annual
endive	*Cichorium endivia*	Asteraceae	annual
escarole	*Cichorium endivia*	Asteraceae	annual

Crop	Species	Family	Life Cycle
fava bean (broad bean)	*Vicia faba*	Fabaceae	annual
frisee	*Cichorium endivia*	Asteraceae	annual
gourd	*Cucurbita* spp.	Cucurbitaceae	annual
gourd (hard-shelled gourd)	*Lagenaria siceraria*	Cucurbitaceae	annual
ground cherry	*Physalis grisea*	Solanaceae	annual
Japanese greens	*Brassica rapa*	Brassicaceae	annual/biennial
kohlrabi	*Brassica oleracea*	Brassicaceae	annual/biennial
lentil	*Lens culinaris*	Fabaceae	annual
lettuce	*Lactuca sativa*	Asteraceae	annual
lima bean	*Phaseolus lunatus*	Fabaceae	annual
melon	*Cucumis melo*	Cucurbitaceae	annual
mustard greens	*Brassica juncea*	Brassicaceae	annual/biennial
oats	*Avena sativa*	Poaceae	annual
okra	*Abelmoschus esculentus*	Malvaceae	annual
orach (mountain spinach)	*Atriplex hortensis*	Amaranthaceae	annual
pea	*Pisum sativum*	Fabaceae	annual
peanut	*Arachis hypogaea*	Fabaceae	annual
pepper	*Capsicum* spp.	Solanaceae	annual
pumpkin	*Cucurbita* spp.	Cucurbitaceae	annual
quinoa	*Chenopodium quinoa*	Amaranthaceae	annual
radish	*Raphanus sativus*	Brassicaceae	annual

Crop	Species	Family	Life Cycle
rice	*Oryza sativa*	Poaceae	annual
runner bean (scarlet runner bean)	*Phaseolus coccineus*	Fabaceae	annual
rye	*Secale cereale*	Poaceae	annual
soybean (edamame)	*Glycine max*	Fabaceae	annual
spinach	*Spinacia oleracea*	Amaranthaceae	annual
squash (summer squash, winter squash)	*Cucurbita* spp.	Cucurbitaceae	annual
sunflower	*Helianthus annuus*	Asteraceae	annual
tepary bean	*Phaseolus acutifolius*	Fabaceae	annual
tomatillo (husk tomato)	*Physalis philadelphica*	Solanaceae	annual
tomato	*Solanum lycopersicum*	Solanaceae	annual
watermelon	*Citrullus lanatus*	Cucurbitaceae	annual
wheat	*Triticum* spp.	Poaceae	annual

Biennials

Crop	Species	Family	Life Cycle
celery	*Apium graveolens*	Apiaceae	biennial
chicory	*Cichorium intybus*	Asteraceae	biennial
Chinese cabbage	*Brassica rapa*	Brassicaceae	annual/biennial
collards	*Brassica oleracea*	Brassicaceae	biennial
Italian dandelion	*Cichorium intybus*	Asteraceae	biennial
Japanese greens	*Brassica rapa*	Brassicaceae	annual/biennial
kale	*Brassica oleracea*	Brassicaceae	biennial
kohlrabi	*Brassica oleracea*	Brassicaceae	annual/biennial
leek	*Allium ampeloprasum*	Amaryllidaceae	biennial
mustard greens	*Brassica juncea*	Brassicaceae	annual/biennial
onion (common onion, multiplier onion)	*Allium cepa*	Amaryllidaceae	biennial
parsley	*Petroselinum crispum*	Apiaceae	biennial
parsnip	*Pastinaca sativa*	Apiaceae	biennial
rutabaga	*Brassica napus*	Brassicaceae	biennial
shallot	*Allium cepa*	Amaryllidaceae	biennial
Siberian kale	*Brassica napus*	Brassicaceae	biennial
smallage	*Apium graveolens*	Apiaceae	biennial
Swiss chard	*Beta vulgaris*	Amaranthaceae	biennial
turnip	*Brassica rapa*	Brassicaceae	biennial

Crop	Species	Family	Life Cycle
beet	*Beta vulgaris*	Amaranthaceae	biennial
Belgian endive	*Cichorium intybus*	Asteraceae	biennial
broccoli	*Brassica oleracea*	Brassicaceae	biennial
broccoli raab	*Brassica rapa*	Brassicaceae	annual/biennial
Brussels sprouts	*Brassica oleracea*	Brassicaceae	biennial
cabbage	*Brassica oleracea*	Brassicaceae	biennial
carrot	*Daucus carota*	Apiaceae	biennial
cauliflower	*Brassica oleracea*	Brassicaceae	biennial
celeriac	*Apium graveolens*	Apiaceae	biennial

Perennials

Crop	Species	Family	Life Cycle
artichoke	*Cynara cardunculus*	Asteraceae	perennial
asparagus	*Asparagus officinalis*	Asparagaceae	perennial
broomcorn	*Sorghum bicolor*	Poaceae	perennial
cardoon	*Cynara cardunculus*	Asteraceae	perennial
chives (common chives)	*Allium schoenoprasum*	Amaryllidaceae	perennial
fennel	*Foeniculum vulgare*	Apiaceae	perennial
sorghum	*Sorghum bicolor*	Poaceae	perennial
wild arugula	*Diplotaxis tenuifolia*	Brassicaceae	perennial

Day Length Sensitivity

Some crops are sensitive to the number of nighttime hours that occur every 24 hours. Flowering will trigger when the nighttime hours reach a certain threshold. Some species can have sensitive and non-sensitive varieties.

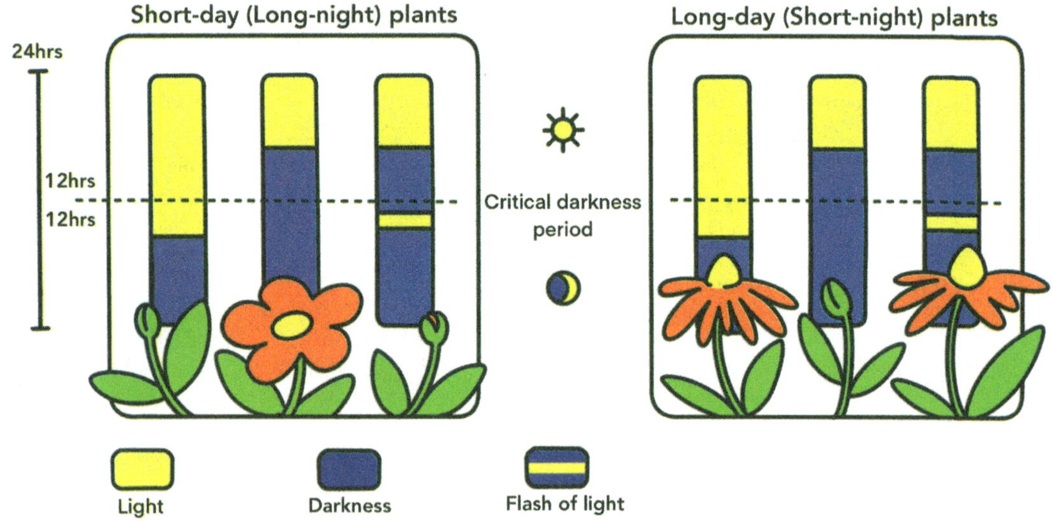

Short- and long- day plant sensativities. Image source: Arielle Price

Short-day (long-night)

Require critical number of hours (usually 14-16) of uninterrupted night (lack of light) to stimulate flowering. Crops such as rice, cotton, soybeans, mung bean, and sorghum are short-day plants.

Long-day (short-night)

Require critical number of hours (usually 14-16) of light to stimulate flowers. Crops such as beets, carrots, lettuce, radish, spinach, and turnips are examples of long-day plants.

Day-neutral

Flowering does not depend on night or day length. Crops such as amaranth, cucumbers, kale, peas, tomatoes and brassicas are examples of day-neutral plants.

Field Planning

Orient and prepare seed production plots so that plants receive uniform sunlight, moisture, and nutrients. Raised beds provide growing space that keeps crop roots out of standing water during rainy seasons. They also allow for targeted irrigation in the dry season, typically through drip irrigation or hand watering. Raised beds without permanent borders will need to be reformed as the edges of the beds fall.

Field Layout:

Plans must be made for the layout of the field. Isolate different varieties of the same crop from each other (in space, time, or with barriers) to prevent cross-pollination, and ensure genetic purity in the seed produced. With varieties that naturally cross-pollinate, larger numbers of plants must be planted to ensure adequate mixing of genetics. Also, consider more than one grow-out plot, which can be helpful if one has problems. Plant enough seeds or transplants to compensate for some plants dying or being intentionally culled. Plan how pollination will occur. Will it happen naturally, or be done by hand to control cross-pollination? Plan for the timing of harvest and processing as well.

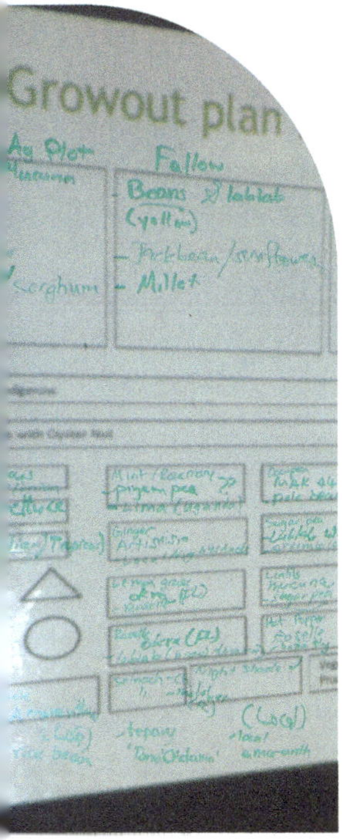

Whiteboard brainstorming seed bank multiplication beds

Beds for growing out seed crops with gravity-fed drip irrigation systems. ECHO Asia, Chiang Mai, Thailand.

Crop Maintenance

A well-grown, healthy crop produces vigorous seeds. Soil should be made fertile and balanced. Slow release of nutrients in the soil is good for seed growth, as an over-abundance of nitrogen can cause plants to put more energy into vegetation than flowers. In the absence of special slow-release NPK fertilizer, the release of nutrients can be evenly spread throughout the growing season by applying small amounts of compost over time. Multiple applications are especially necessary in high rainfall areas.

Ideal plant spacing for seed production allows for the development of mature seed. For crops that produce seed in fruit, such as tomato, space the plants as you would for harvesting fruits. For leafy crops and root crops, such as lettuce and carrot, which are not fruit-bearing, plants with mature seed might be larger than those harvested for eating. Consider staking or trellising large plants, which helps prevent plant diseases by keeping foliage and fruit off the ground and increasing airflow within the crop canopy. Manage and reduce disease and pests as much as possible. Seed from a diseased or infested plant will likely be contaminated and spread the disease or pest. Manage weeds to increase yield and reduce disease risk (Organic Seed Alliance, 2010). During dry periods, irrigate the crop to meet crop water requirements and prevent wilting.

Help multiplying seed

If you do not have sufficient space or the right microclimate for growing some or all of the seeds that you need, consider contracting or working with local farmers to multiply seeds of crops. To ensure production of pest- and disease-free, quality seeds, it is ideal to have farmers grow seeds of crops that they are familiar with and know how to manage. Visit farmers during the production season to help monitor crops and provide assistance as needed. Also, consider sourcing seeds from local markets or other trusted suppliers in the region.

Chapter 6. Plant Pollination

Pollination and Fertilization

Pollination and fertilization are two biological processes that must happen in order for seeds to be produced. **Pollination** occurs as pollen grains produced by stamens (male flower structure) are transferred to the stigma. Upon reaching an ovule, **fertilization** happens as a sperm cell (male reproductive cell) from a pollen grain unites with the female reproductive cell (egg) in the ovule. Fertilization depends on successful pollination, the focus of this chapter.

Self-Pollination and Cross-Pollination

An understanding of pollen transfer helps seed bank personnel decide how best to multiply seeds of a given plant. In this regard, we will distinguish between crops that are self- versus cross-pollinated. **Self-pollinated** plants have complete flowers, meaning that each flower has male and female structures. Moreover, within a flower, these structures will mature at the same time and in a way that allows the pollen from an anther to easily reach the stigma. Thus, pollination in self-pollinated crops can occur within an individual flower, without the need for a vector (insects or wind) to transfer the pollen.

Cross-pollinated crops, on the other hand, rely on insects or wind to transfer pollen between flowers. Practices that support pollinators such as bees will improve seed production of cross-pollinated crops. In an *ECHO Technical Note*, Motis (2010) added the following:

A male (left) and female (right) cucurbit flower.

Cross-pollinated crops may or may not have complete flowers. Those without complete flowers (and some with) will have male and female flowers. Some are monoecious, with male and female flowers occurring on the same plant. A prime example of this is the cucurbit family, with cucumber, squash and pumpkins, the vines of which have a number of flowers with each flower either male or female. Other cross-pollinated crops are dioecious, with male and female flowers occurring on different plants. Dioecious crops, including yams and asparagus, require both male and female plants to produce viable seed. The stamens and pistils of cross-pollinated crops often do not mature at the same time.

Examples of self- and cross-pollinated crops are listed in the table below. There are a few complexities that, once understood, will minimize confusion. These are listed below:

- Pollen transfer in self-pollinated crops occurs mostly within individual flowers; however, there is often some crossing that occurs between plants due to insect activity.
- Transfer of pollen between different flowers on the same plant can occur unless the plant has a way to prevent that from happening. From a genetic perspective, this is a form of self-pollination even though crossing has occurred. This is because both parents are genetically the same, since both flowers are on the same plant.
- Crossing of pollen between plants can be done by humans, for both self- and cross-pollinated crops.

Application of this knowledge is presented in Chapter 8 on maintaining seed purity.

Crop				
Common name	Latin Name	Self-pollinated	Cross-pollinated	Wind/insect (if cross-pollinated)
Bean	*Phaseolus* spp. *Vigna* spp.	X		
Lettuce	*Lactuca sativa*	X		
Pea	*Pisum sativum*	X		
Tomato	*Lycopersicon esculentum*	X		
Corn	Zea mays		X	Wind

Crop				
Common name	Latin Name	Self-pollinated	Cross-pollinated	Wind/insect (if cross-pollinated)
Beet	*Beta vulgaris*		X	Wind
Spinach	*Spinacia oleracea*		X	Wind
Swiss chard			X	Wind
Crucifers (e.g. cabbage, collards)	*Brassica* spp.		X	Insect
Cucurbits (e.g cucumber, squash, pumpkin)	*Cucumis, Cucurbita,* and *Luffa* spp.		X	Insect
Okra	Abelmoschus esculentus		X	Insect
Roselle	*Hibiscus sabdariffa*		X	Insect
Onion	*Allium cepa*		X	Insect or wind
Carrot	*Daucus carota*		X	Insect or wind
Parsley	*Petroselinum crispum*		X	Insect
Sunflower	*Helianthus annuus*		X	Insect

Open-Pollinated (OP), Hybrid, or GMO?

Open-Pollinated

Open-pollinated varieties (OPVs) are the result of unrestricted pollen transfer. Seeds saved from OPVs produce offspring similar to their parents. Heirlooms are OPVs passed down from generation to generation of farmers and gardeners within a community. Farmers have selected seed from plants displaying desired traits for centuries.

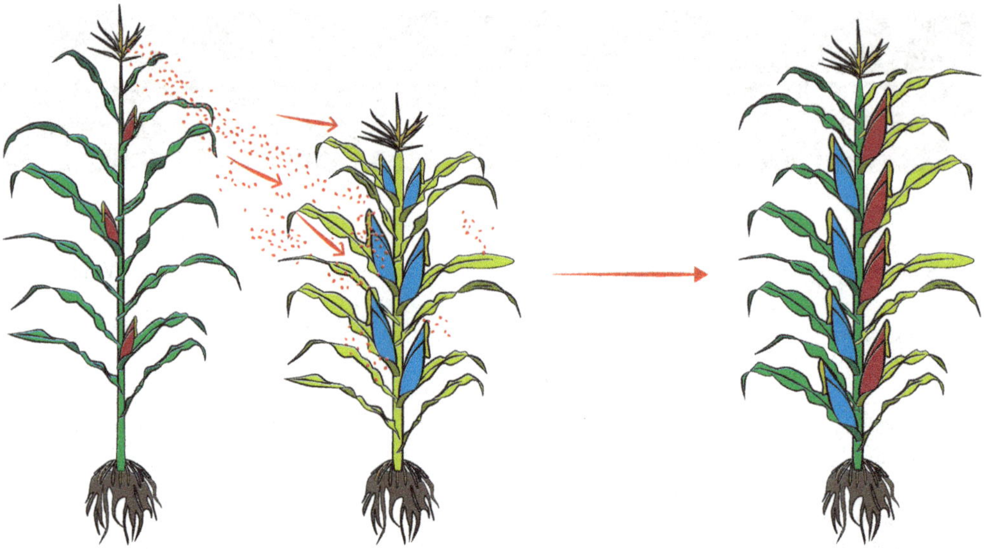

Controlled crossing of parent populations to create a hybrid. Image source: Andre Fonseca

Hybrid

Commercial **hybrid seeds** are the result of carefully controlled pollination (crosses between specific parent populations) to produce plants with desired traits and high vigor also known as heterosis (Figure above). An example of a hybrid is the controlled cross of a variety of corn resistant to drought with a variety of corn that scavenges nitrogen in the soil. The hybrid crop is the first generation (F1) cross of these two separate varieties and has the desirable traits of both. Hybrid crops have increased yield, resilience, or adaptability than either of the original varieties. This is called hybrid vigor and is very advantageous for crop production within one growing season. The main disadvantage of hybrid seed is that the seeds saved from hybrid plants (F2) may produce plants with less desirable traits than their hybrid parents. The desirable traits conferred to the first generation (F1) are not consistently found in the subsequent generations (F2). Therefore, seed from hybrid crops (F1) should not be saved for future cropping seasons.

Loss and inconsistent expression of hybrid traits in F2 generation. Image source: Arielle Price

Creolization

Creolization of a crop occurs when farmers continue to save and plant seed from an improved variety (hybrid or OP), or an improved variety regularly crossed with landraces (local varieties), for 4-15 years. A creolized crop's genetic traits stabilize over time, and after 15 years of creolization scientists consider the resulting crop a new landrace. Creolized crops may retain beneficial traits from their improved variety parent, with the added advantage of acclimatization to the growing region (Bellon et al., 2003).

What is a Genetically Modified Organism (GMO)?

Genetically modified organisms, including plants, have had DNA changes in their genome in order to confer some trait or outcome. Horizontal gene transfer happens when viruses, bacteria, fungi, or other vectors exchange genes with a species. Vertical gene transfer happens when genetic changes occur between generations in a species. The stability of the genetic change depends on many different factors. Farmers may have access to GMO seeds from local agrodealers, but most small-scall seed banks do not have the infrastructure required for gene transfer technology. For asexually propagated species such as sweet potato, GMO plants can be grown on a small plot near the seed bank for the selling/sharing of cuttings/air layers.

Classification	Advantages	Disadvantages
Open Pollinated	Seeds grow 'true-to-type'	May only be adapted to local climate
	Can save seed year after year	Requires maintenance of genetic purity
	Affordable and accessible	
Hybrid	Improved characteristics..(e.g. disease resistance, high yield)	Seeds cannot be saved from one cropping season to the next without inconsistencies in trait expression
		Expensive
		May not be available
Genetically Modified	Improved characteristics..(e.g. disease resistance, high yield, etc..)	Expensive
		May not be socially accepted

Chapter 7. Conducting Variety Trials

The Importance of Variety Trials

Within a species of plants there are usually types of plants, referred to as varieties, that have distinct, reproducible traits. A specific tomato variety, 'Homestead 24' for example, is known for its heat tolerance. Crop varieties include naturally-adapted local types of plants called landraces as well as cultivated varieties called cultivars. When starting a CSB look for promising varieties by observing what farmers are growing, visiting local markets, and finding out what is available in seed stores or through NGOs in the region.

Every variety has advantages and disadvantages, and CSBs can play a helpful role in determining and communicating what they are. CSBs have the ability to grow out multiple varieties and match them appropriately to their community's needs.

All variety claims should be tested for the local context. A corn variety that is said to be an "improved" high- yielding variety could actually be less productive than local varieties under local conditions without irrigation or high amounts of fertilizer. Be conservative when promoting a new variety, even if there is only a small chance it could fail. Any failure is potentially detrimental to a smallholder farmer and could cause local partners to lose trust in future recommendations. Start with small-scale trials on CSB property. To further reduce risk when introducing new varieties to farmers, encourage them to start with a small planting and expand by small increments in following seasons if results are favorable.

Crop Characteristics to Select For

When looking for promising varieties to evaluate, identify traits that are important to farmers and gardeners served by the CSB. What metrics best determine the performance of a given variety? Often, growers focus solely on yield. That is an important metric, but not the only one.

Local varieties have the advantage of familiarity. Traditional farming methods developed alongside traditional varieties. Local flavor preferences, methods of culinary preparation, and means of storage are all factors that may have influenced the selection for a variety over time. The crop traits a community desires do not always match the crop traits commercial seed producers focus on. For example, some native Peruvian farmers use corn leaves as fodder for their livestock and corn stalks as fencing material. The importance of the leaves and stalks would be missed by evaluating varieties based solely on yield.

The book, *Two Ears of Corn*, documents a case where a new corn variety was introduced to a community. It had twice the output of the native variety with similar inputs. However, the texture of the corn was unsuitable for their traditional cooking methods (Bunch, 1982).

These examples show why there is often resistance to new varieties. To help address this, consult communities to find out which traits they look for in a crop. Take note of what they do with every part of the plant and see if the same things can be done with the varieties you trial.

How to Organize a Variety Trial

The best way to test how well varieties will grow under your conditions is to grow them in your context. Grow several varieties side-by-side and compare their traits to help narrow down which varieties you want to use.

Try to develop a sense of detachment from any one variety. If you believe one variety is better from the start, you might unwittingly give that variety extra care or attention. This could make it perform better and ruin your results.

Consistency is Key

Be scientific in your trial design and implementation. Keep as many variables consistent as possible to obtain the most accurate results. Take detailed notes throughout the season and be methodical and consistent with trial establishment, trial maintenance, and data collection. When selecting growing plots, ensure they receive similar amounts of sunlight and have similar soil. Provide consistency of care in pruning, weeding, pest management, etc. If a plot or plant receives slightly different care, it will be difficult to discern the true cause of any perceived varietal differences. However, if the only variable that changes is the variety, differences in performance can be attributed to the variety. If one variety gets a certain amount of fertilizer or water, give the others the same amount.

Representative Plots

It is not necessary to plant a whole field for a variety trial, but at a minimum, you should have ten to twelve plants of any given variety for a meaningful assessment. Plant them in plots established in close proximity to each other to increase the likelihood of similar growing conditions. Select a location for the trial that will allow you to evaluate the varieties in plots that are representative of the client's growing conditions. For example, if the crop is typically grown in a sloped upland setting with poor soil, consider conducting your trial on a hillside with poor soil.

A simple trial could look like the following illustration:

Test Variety 1	Test Variety 2	Test Variety 3	Standard/Check Variety

Note: The standard variety acts as a control, since you are trying to determine how the new varieties compare to the current one.

Replication

While a simple trial design can be informative, a more scientific, complex trial design provides opportunity for greater accuracy and more confidence in the results. Different sections of the field may have different pest pressures or receive different amounts of water. Conditions might vary from year to year. You can account for that variability by repeating the test in different areas of the field over a few seasons or years.

A replicated design in a field could look like this replicated design for trialing three varieties:

Block A	Test Variety 1	Test Variety 3	Test Variety 2	Standard/Check Variety	Gradient of external factor (sun-light, slope, etc.) ↓
Block B	Test Variety 2	Standard/Check Variety	Test Variety 3	Test Variety 1	
Block C	Standard/Check Variety	Test Variety 2	Test Variety 1	Test Variety 3	

This is known as **Randomized Complete Block Design** (RCBD). Each replication is a "block" of plots (in the figure above, blocks are lettered and plots are numbered), and every block represents a different area of the field. In this example, the blocks are arranged so that environmental differences, such as elevation, are consistent across the block. Each block contains one plot of each variety. Varieties are randomly assigned to plots within each block. Randomization helps reduce influence from bias.

Evaluation Throughout the Trial

Take careful notes during the trial. Do some varieties have higher pest pressure at different times? Do any varieties show signs of needing more fertilizer, or needing it at different times? Do they flower at different times? What is the growth habit of each variety? Pay close attention to metrics that your community is interested in. Ask local farmers to look at the varieties during the trial. Their input is invaluable.

Since there are so many variables to consider, it may be beneficial to use a form when taking notes. An example form is provided on the next page, which provides examples of metrics to measure

Variety Trial Name:
Genus/Species: **Date:**

Methodology, Timing, and Plant Height Data										
Plant Identification				Dates					Measurements	
Assigned Number	Replication	Accession	Variety Name	Seeding mm/dd/yy	Planting (for transplants) mm/dd/yy	Flowering mm/dd/yy	Maturity mm/dd/yy	Final Harvest mm/dd/yy	Plant height at maturity cm	Plant spacing in and between rows cm

Plant Identification		Pest Observations Data to be Recorded at Least Twice (e.g. at flowering and fruit set)				Harvest				
		Leaf and stem damage severity (scale of 0 to 5)				Date	Plant Part	# of Plants Harvested	Yield	
#	Date mm/dd/yy	Insect (identify)	Nematode	Other (specify bacteria, fungus, virus, etc.)	Observations/Comments	mm/dd/yy	(fruit, leaf, whole plant, root, etc.)	(for each accession in each rep)	Weight of part (g or kg)	# of parts harvested

Evaluation after the Trial

After the field portion of the trial is complete, there is more data to collect (depending on your chosen metrics). You can evaluate factors like final yield, flavor, texture, and how well the harvested products can be stored and transported. Invite local farmers and consumers to take part in a variety taste test or other forms of assessment. Keep local context in mind when evaluating.

To evaluate taste, you can use something as simple as the form below. Taste is subjective, with preferences varying between individuals and influenced by culture. Nonetheless, findings can be insightful, especially when combined with other factors such as overall ratings by farmers.

Please rate the oranges below *5 = top choice and 1 = last choice.*

Hamlin	5	4	3	2	1
Ambersweet	5	4	3	2	1
Parson Brown	5	4	3	2	1
Pineapple Orange	5	4	3	2	1
Varn	5	4	3	2	1

A form like the example characteristic preference form below could be used to help determine what a community finds desirable in an orange.

Citrus Score Card

Date:________

Citrus Type:______________________________Cultivar:______________________________

Circle one:

Season:	Early	Mid	Late
Seediness:	Low	Medium	High
Juiciness:	Low	Medium	High
Ease of Peel:	Easy	Medium	Difficult
Sectioning:	Poor	Good	Great
Overall Appeal:	Poor	Good	Great

If you trialed many varieties but did not replicate any, consider selecting the most promising varieties for replicated trials.

The following table was part of an ECHO Annual Leafy Greens Variety trial by VanNocker and Reed (2012) evaluating productive potential of different leafy annuals.

Table 1. Fresh weight of leaves or heads for ten different leafy green crop varieties. Data were averaged over three replications.

Crop		Weight (g/plot) of leaves harvested at varying times (days) after seedi			
Latin name	Variety name	57 days	64 days	78 days	Total
Brassica juncea	Thai Mustard Green	425	1987	565	2976
Brassica juncea	Japanese Giant Red	441	995	2337	3773
Brassica juncea var. *tumida*	Big Stem Mustard	71	203	930	1205
Brassica oleracea	Toscano Kale	–	186	1698	1884
Brassica oleracea var. *alboglabra*	Gailan	–	345	1811	2156
Brassica oleracea var. *narinosa*	Vitamin Green	1240	1203	3786	6228
Brassica oleracea var. *perviridis*	Summerfest Komatsuna	734	1260	1576	3570
Brassica oleracea var. *rosularis*	Hon Tsai Tai	492	522	2453	3468
Brassica oleracea subsp. *Nipposinica*	Purple Mizuna	636	713	876	2225
Chrysanthemum coronarium	Large Leaf Tong Ho	–	352	668	1020
P value[Z]		<0.001	<0.001	<0.001	<0.001
LSD value[Y]		416	460	811	889

[Z]Within each column, at least two means are statistically different because the corresponding P value is ≤ 0.05.

[Y]A least significant difference (LSD) value was calculated for each column. Any two means within a column are statistically significant if

Only one variable was measured, so the data collected on that variable is extensive.

Note that production at set intervals and overall were both measured, and statistical analysis was done to account for variation. These steps strengthen the data but may be more than what is necessary for some contexts.

Varieties of okra (top), pumpkin (middle), and corn (bottom)

Cowpea vines at the end of a 12-week trial

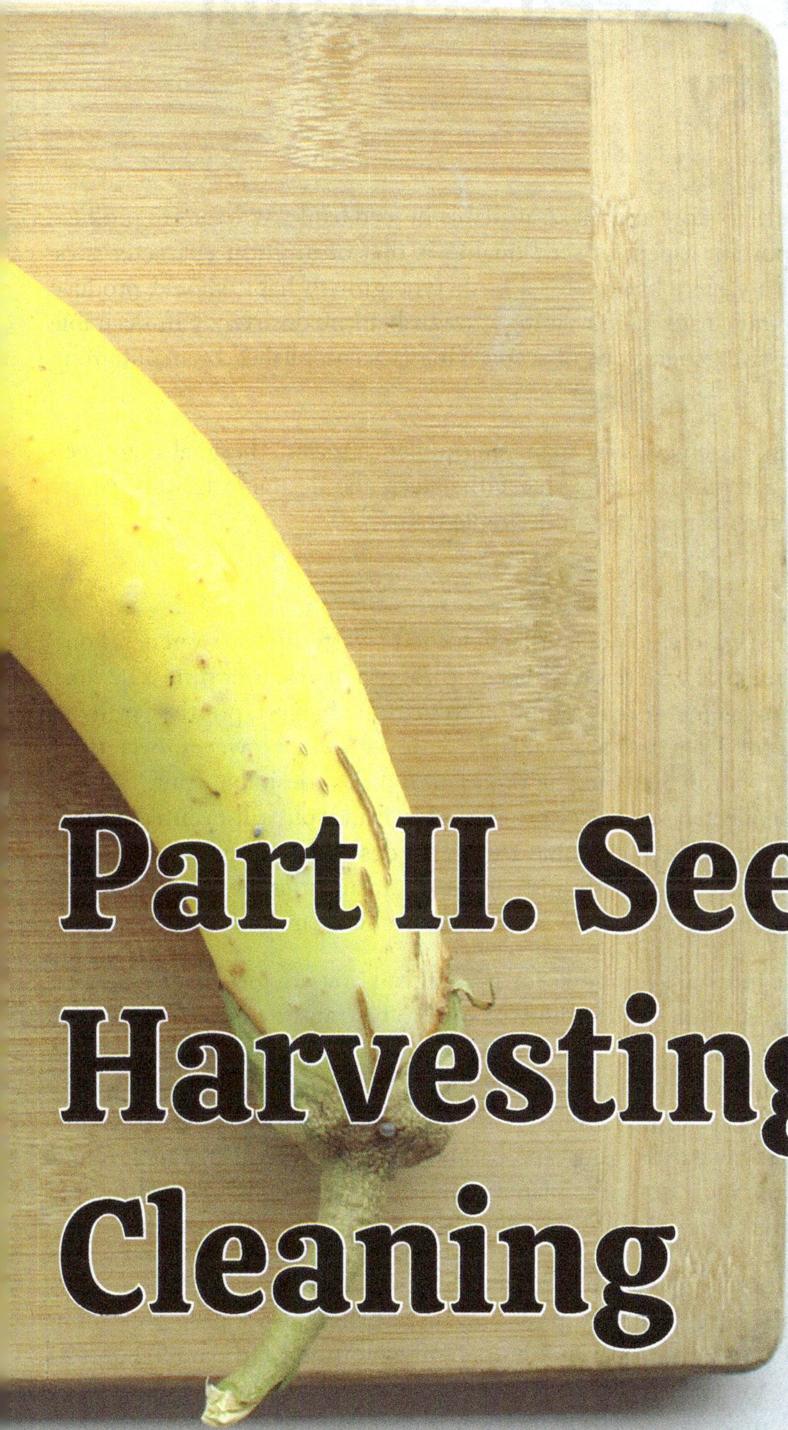

Part II. Seed Harvesting & Cleaning

Chapter 8. Seed Selection and Purity

When multiplying seeds to store and make available in seed banks, we want to produce quality seeds. An important aspect of seed quality is that of uniform characteristics. For example, seeds of a variety known for a bush-type growth habit should produce plants that have that trait instead of a vining growth habit. Consistency in desirable plant traits over successive generations of seed saving is accomplished by maintaining genetic purity.

Genetic purity, in turn, is achieved by managing pollen transfer. The goal is to keep plants from crossing with other plants that either lack the desirable trait(s) we are interested in maintaining, or that have undesirable traits.

Selection

After you select a desired variety, do not let that work go to waste after a few generations. Often, part of the reason local varieties have become less productive over time is because of poor seed purity management. It is tempting to take the biggest, prettiest fruits to market and collect seeds from the less valuable, leftover fruit. However, over time, saving seed from inferior fruit leads to the production of less desirable fruit. Because you will have always collected seeds from undesirable fruit, you will eventually have seeds that produce plants with undesirable fruit. This describes an agricultural term: **Seed selection.** In this case, seeds with undesirable characteristics have been selected and saved. Your aim should be to select seeds with desirable characteristics.

First, determine what traits are desirable for the cultivar you are growing. It may be fruit size, fruit taste, fruit color, or the ability of fruit to be transported to market (See the table on the next page for more examples). Once you know what traits to look for, walk through your production plot to examine individual plants. When you identify which plants have the most desirable traits, mark them to show they are for seed (not to be eaten or sold). You can use flagging tape, bamboo poles with flagging tape, ground flags, or other easily seen indicators. The plants selected for seed should be healthy, free of pests, and toward the center of a plot to reduce the risk of cross-pollination.

The selection of high-quality seed is essential. Save high quality seeds from high quality fruit. Pure lineage, resistance to pests and disease, high yield, proper cleaning and drying, and ideal storage (appropriate temperature, moisture, and oxygen control) are factors that allow for healthy future plantings. There is no point in saving poor quality seeds! Instead, make your seeds better over time (continual improvement).

Selecting for Seed with Improved Characteristics

The following table gives some examples of selection characteristics used to identify high quality seeds to harvest, though of course, your needs may vary. For more details about specific crops, search for them on ECHOCommunity.org.

Sweet Corn	Early producers are desirable—mark plants that produce cobs early
Crops that go to seed after the edible stage (lettuce, cabbage)	A long edible stage is desirable—save seeds from plants that are slow to produce a seed head.
Crops with multiple pickings (beans)	Collect seed from the best plants as they mature.
Root crops (carrots)	The shape of the root often impacts its market value—select plants with the best-shaped roots (long/thin- sandy soils; short/stubby- clay soils).
Tomatoes	Fruits need to be shaded from the sun and protected from the onset of disease—collect seeds from plants with good leaf biomass, early fruit production, and delicious fruit that does not crack or split.

Knowing When to Harvest for Seed

Every crop needs to be harvested at a different point in its lifecycle for the highest quality seeds. Often, the optimal time to harvest for seeds is different from the optimal time to harvest for consumption. The following table gives an example of the optimal time to harvest a few crops for seed production:

Cucumbers	Pick after the edible stage. Wait for the cucumbers to change color and/or shape, depending on variety.
Squash	Harvest seed squash at the same time as the rest of your crop.
Crops with multiple harvests (beans)	Note what color the beans turn as they develop. Typically, ripe beans get very dark in color. In that case, only pick the ones for seed that are the darkest in color.

Root crops (carrots)	After harvest, leave some plants in the ground. They will continue to mature and produce seeds even as the root becomes hard and inedible.
Tomatoes	Wait as long as possible to pick the fruit. The seeds are at the optimal stage when a ripe fruit first shows sign of rotting.

Seed Purity

Seed purity refers to the uniformity in plant characteristics or traits of a seed lot as it is planted over time. Remember the previously mentioned concepts regarding inbreeding and outcrossing crops, as those concepts will help you determine the number of plants needed to save seeds from to maintain desired traits. Special care must be taken to isolate different varieties of the same crop from each other (in space, time, or with barriers) to prevent cross-pollination, and ensure genetic purity in the seed produced. With varieties that naturally cross-pollinate such as corn, larger numbers of plants must be planted to ensure adequate mixing of genetics. Plan how pollination will occur. Will it happen naturally, or be done by hand to control cross-pollination?

If growing out more than one variety of a crop, label the locations in the field. Mark plots with labeled stakes, and record plot locations on paper or electronically. This way, if plot stakes are pulled up or lost during weeding, there is a written plot plan/map to refer to. As a general rule, it is not wise to multiply seeds harvested from variety trials. In these trials, there is a high potential for crossing, as the various varieties are grown in close proximity to each other. Instead, go back to an original source of seeds for the varieties you want to plant out. Before planting preliminary variety trials, it may be wise to keep some seeds back for potential multiplication purposes.

When saving seed of multiple varieties of a cross-pollinated crop, the varieties must be separated to prevent the mixing of pollen between varieties. This can be done by planting varieties an appropriate distance away from each other (see the Appendix).

Seed purity, when planting multiple varieties of a cross-pollinated crop, can also be maintained by varying the planting time of each variety so that they are not all flowering at the same time. This is most applicable if there is irrigation during the dry season. Another possibility is to cage the varieties, using hoops with row-cover cloth or fine-meshed screen draped over them, to prevent insects from transferring pollen between varieties.

Bamboo supports for exclusion structure

Exclusion structure net (the final step is to bury the base of the netting around the structure)

Roguing for Seed Purity

There is another way to maintain the quality of your seeds, which becomes especially important if other varieties of the same crop are being grown nearby: **Roguing.** That means pulling out any "rogue" (different) plants that sprout up from your good seeds. Because of genetics, plants will sometimes sprout that are significantly different from the parent variety. The chances of this become higher if different varieties are grown nearby, potentially contaminating your variety with the genes of the others. Genetics could also lead to seedlings that simply are not as vigorous or healthy. It's important to pull these out as soon as possible. Otherwise, you not only waste space with a plant that's less productive, you also risk passing on its bad genes to the other plants in your field. Before pollination, rogue plants that are diseased, under-performing, mutated, infested, or less desirable. It may also be possible to remove male flower structures (e.g. maize tassels) of undesirable plants before the pollen is viable.

X indicating plants that would be rogued. Image source: Arielle Price

Varieties never stay static; they will always change. Your job is to stay vigilant and make sure they change for the better!

Chapter 9. Seed Cleaning Techniques

Seed cleaning workshop session in Pinlaung, Myanmar, 2020.

Why Proper Seed Cleaning is Important

Seed cleaning is one of the most important steps in the process of saving seeds. Proper care at this stage is critical to the storage longevity of your seeds.

Note: Some seeds will need to be processed and cleaned before they can be dried, while other seeds may require an initial drying period before they can be cleaned. For this reason, we will cover the topic of seed cleaning first, followed by a chapter on seed drying.

The primary reason to properly clean seeds before drying and storage is that clean seeds prevent pests and diseases from multiplying during storage. It is critical to remove debris, in which insects may lay their eggs, and limit any opportunity for the multiplication of fungal, bacterial, and viral diseases during storage. Proper cleaning of seeds will also reduce the volume of seed material in storage containers, which is important for smaller seed banks limited in available storage space.

Separation of undesirable materials, including:

- Gravel
- Organic matter and other debris
- Weed seeds
- Immature (underdeveloped) seeds
- Cracked or broken seeds
- Moldy or diseased seeds
- Seeds damaged by pests

Seed Cleaning Techniques

Different crops and quantities require different techniques and equipment for cleaning.

If seed samples are small, proper cleaning can be completed by hand using a combination of baskets, screens, sieves, tweezers, and other household items. For larger batches, it is recommended to use machinery capable of separating and cleaning debris of different sizes.

Releasing the Seed

Threshing is the action of breaking up plant material to expose or release seeds. This can be done by crushing, rubbing, beating, stomping, or grinding the plant material.

A food processor or blender can be used to cut fruit to expose seed. Test a small batch first to make sure the blades do not cut the seeds. You may need to add water to the blender. The seeds can then be winnowed (dried chilis), or float tested (tomatillos, eggplant) to remove the flesh and unviable seeds that float. See winnowing and float testing below.

Dried bean pods or seed heads can be placed in a woven bag and beaten to release the seeds. This technique works well if the plant material is very dry when placed in the bag. At ECHO, we do this for pigeon pea, cowpea, velvet bean, Lagos spinach, sunn hemp, chipilin, Sesbania, grain amaranth, and vegetable amaranth. The seeds are then separated from the large plant material using a sieve and then winnowed.

A burr grinder is an option for separating small seeds from grain heads. The plates should be adjusted so that the seeds are released but not crushed.

A bucket or barrel thresher can be made for low cost. This utilizes a metal rod with chains put into a bucket or barrel through the lid. The metal rod is then attached to a drill. The chains spin around and beat the seeds out of the plant material. The plant material must be very dry first for this to be effective.

There are many different small-scale grain threshing machines for small grains. Some use a pedal (e.g. bicycle threshers) or hand power, and some can be modified to use a motor.

Fermenting is the action of using bacteria to break down the gel or flesh surrounding some seeds. This can be used for crops like tomato, passionfruit, Malabar spinach, neem, and cucumber. See fermentation method below.

Removing the Refuse

Float testing is the action of separating viable seeds from unviable seeds using water. This does not work for every type of seed but does work well for eggplant and tomatillo seeds.

Be sure to separate the seeds from the flesh first, then mix the seeds with water and pour off anything that floats. Keep the seeds that sink and dry them on a tray in a shaded area.

Float tests can also be used for some beans and grains. Dry the seed thoroughly first and then put a small batch of seed in a container with water and stir. Pour off anything that floats and keep what sinks. Quickly dry the good seeds again to prevent imbibition.

Winnowing is the action of separating plant material from seeds using a fan or light wind. It is best to first remove larger debris from seeds using a sieve in order for winnowing to be effective. Wear protective eyewear, gloves, and a face mask when winnowing hot pepper seeds as the debris can be irritating.

Pour seeds from one container into another in front of a fan to remove lighter debris. Do a test first to adjust the fan speed to prevent losing good seed.

Wind can also be used to winnow seed. Test a small sample first, to make sure the wind does not carry the good seed away with the debris.

Seed can also be separated by using a pan and tossing seed up and around to move the lighter debris to the higher end of the pan. The lighter debris can be blown off.

Winnowing machines are expensive but can quickly remove debris and unviable seeds.

Hand sorting is the action of removing damaged or unviable seeds by hand. If the above methods do not work to remove plant material or damaged, discolored seeds then hand sorting is a proven method. It is time consuming but a great community builder.

Seed Cleaning Methods

While there are many seed cleaning methods used for a variety of different crops, we will highlight some of the most widely used techniques, focusing on methods that can be done by hand using locally available resources. Cleaning techniques have been broken into three main categories: dry, wet, and fermentation methods.

Dry Method

Thai Basil (*Ocimum tenuiflorum*)

Step 1. After allowing basil plants to go to seed, select mature brown seed heads for harvest. Green seed heads are immature and will not yield viable seeds.

Step 2. Separate seed pods from the stem by hand.

Step 3. Place seed pods into a mortar and gently pound seeds to separate them from their pods. Be careful not to pound too hard, or seeds may be cracked and destroyed.

Step 4. Once seed pods have been crushed, separate seeds using different-sized sieves. The first screen should be big enough for all the seeds to fall through while trapping the majority of the pods.

Step 5. Once the majority of the seed pods have been separated from the seeds, use a smaller screen to further separate the seeds and any remaining debris. This time, seeds will remain while debris is pushed through the sieve.

Step 6. Repeat step 5 using an even smaller screen to massage and push through any remaining debris. Clean seeds should remain in the sieve.

Step 7. Lay seeds on a tray and manually separate low-quality seeds. Broken, discolored, and immature seeds are pushed to the side.

Step 8. Though tedious and time-consuming, a set of tweezers is useful to manually pick out any remaining low-quality seeds.

Step 9. The final product (left) contains seeds of high quality that are uniform in size and maturity. These seeds are ready to be dried further and stored.

Wet Method

Eggplant (*Solanum melongena*)

Step 1. To harvest eggplant for mature seeds, pick the fruit later than usual, waiting longer than the harvest time for use as a vegetable. In this case, the eggplant on the right is ready to be harvested as a vegetable, while the eggplant on the left is older and ready to be harvested for seed.

Step 2. Place the eggplant on a cutting board and gently pound the fruit on all sides with a pestle to loosen the seeds within.

Step 3. Use a sharp knife to slice open the fruit from top to bottom.

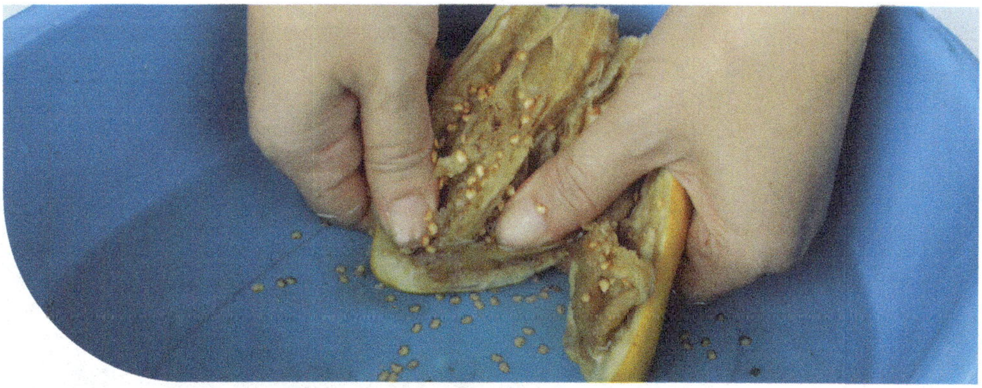

Step 4. Gently massage the fruit while scraping its seeds into a bucket of clean water. Seeds should easily separate from the flesh following step 3.

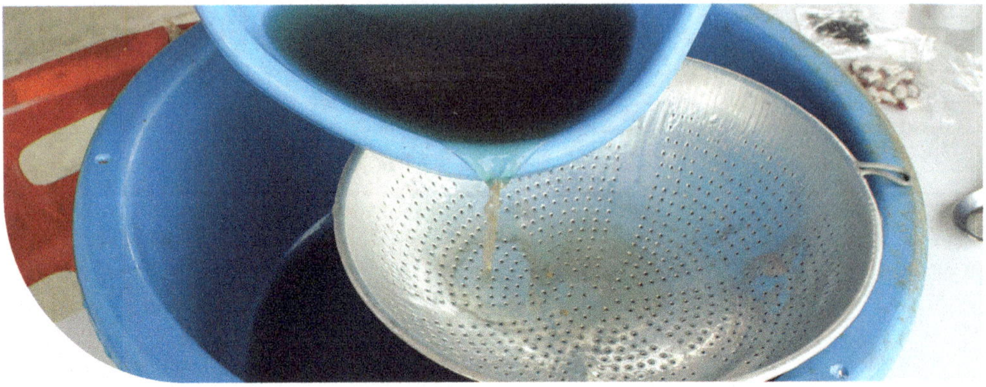

Step 5. Allow the seeds to sink to the bottom and carefully pour the water from one container to another through a sieve to separate the seeds.

Step 6. Seeds sink to the bottom of the first container and remain there while pouring. The sieve collects any seeds that escape along with the flesh and debris and can be picked out by hand.

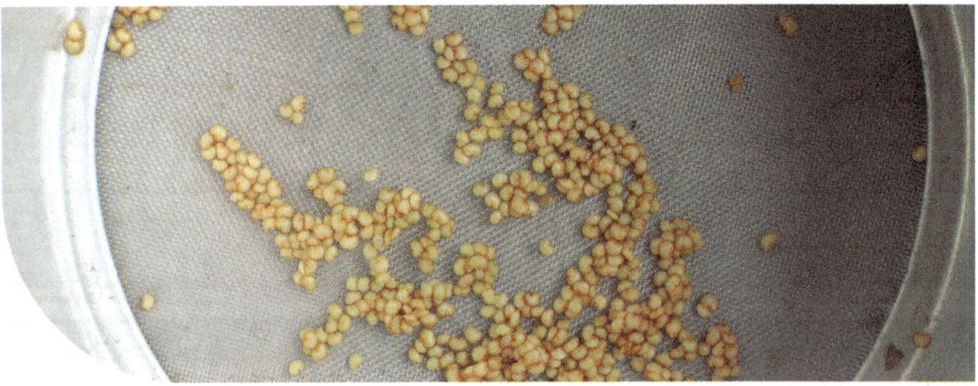

Step 7. Transfer these seeds into another sieve and thoroughly rinse them using clean water.

Step 8. Manually remove any broken, discolored, or immature seeds. Seeds are now clean and ready to be dried, then stored.

Papaya (*Carica papaya*)

Step 1. To harvest papaya for mature seeds, pick the papaya fruit later than usual, waiting until the fruit has changed color entirely and feels soft to the touch. In this case, the lower papaya is ready to be harvested for its seed.

Step 2. Slice open the fruit from top to bottom to access the seeds.

Step 3. Use a spoon to scoop out the seeds into a small piece of mesh. Note: The remaining fruit may still be useful.

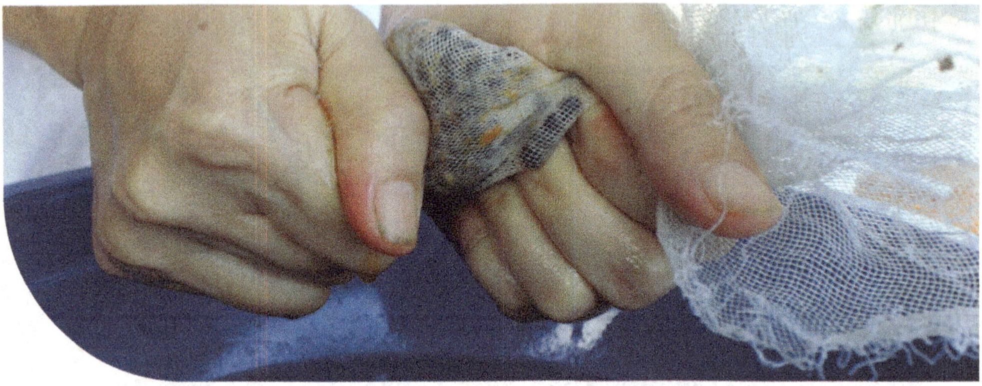

Step 4. Using the mesh around the seeds, carefully massage the contents to separate the flesh from the seeds.

Step 5. Once the seeds have been thoroughly massaged using the mesh, empty the contents in a bucket of water, then strain through a sieve.

Step 6. Rinse seeds repeatedly to separate the remaining flesh from the seeds.

Step 7. Note the difference between seeds that have been cleaned (right) and those that have not (left).

Step 8. Papaya seeds are clean when all the flesh encasing them has been removed.

Fermentation Method

Tomato (*Solanum lycopersicum* syn. *Lycopersicon esculentum*)

Step 1. Collect only mature ripe fruit for seed saving. Green fruits will yield immature seeds and are not suitable for reproduction. Cut fruit lengthwise from top to bottom.

Step 2. Using a spoon, scoop the seeds into a container of clean water.

Step 3. Use the edge of the spoon to gently scrape away any seeds that remain attached to the fruit. Note: The remaining fruit may still be useful.

Step 4. Once all the seeds have been collected, cover and allow them to sit overnight in the container of water. Within hours the natural process of fermentation will begin, breaking down the fleshy material attached to the seeds. This process will allow for easy cleaning of seeds.

Step 5. After 24 hours of fermentation, carefully pour out the liquid, leaving the seeds in the container. A sieve can be used to catch any seeds that escape.

Step 6. The seeds remaining in the container are mostly clean at this stage. Pick out any seeds trapped in the sieve and add them to the container.

Step 7. Add additional clean water to the mostly clean seeds, stir the water, and repeat step 5. This step should result in seeds that are almost completely separated from the flesh of the fruit.

Step 8. Transfer seeds into a sieve and run under water with pressure to remove any remaining flesh.

Step 9. Seeds are completely cleaned and ready to be dried, then stored. Use a screen to dry seeds and provide airflow. Seeds may stick together; this is normal.

Part III. Seed Storage

Chapter 10. Seed Drying

Traditional drying and storage methods for conserving seeds in Laos

Traditional drying and storage methods fo conserving seeds in Cambodia.

Importance of Seed Drying

Adapted from: *ECHO Best Practice Note* 5 (ECHO Staff, 2016)

This chapter will focus on orthodox seeds because they can be stored under dry conditions. They represent grain crops and most vegetables.

Drying seeds is critical to the overall success of a CSB, as seeds with a high moisture content do not remain viable for very long in storage. Moisture favors mold and life processes that lead to premature germination and the depletion of food reserves that seeds need to grow when planted. Newly harvested seeds typically retain field moisture from dew or rain. They may also be high in moisture if harvested a little early to avoid pest damage. There is a need, therefore, to ensure that seeds are dry enough to store, the focus of this chapter. Once placed into storage, it is also important to keep seeds dry, which will be discussed in Chapter 11 on seed storage technologies.

Drying seeds in the humid tropics is challenging, with high relative humidity and constant cloud cover during the rainy season. Below we will explain principles for drying seeds without injuring them and present options for both natural and artificial drying approaches. Natural drying requires no electricity or moving parts, making it the most inexpensive approach. Disadvantages of natural drying methods are the lack of precise control over temperature, air circulation, and humidity. As a result, drying is slower on cloudy, rainy, or cool days than on warm, sunny days. Artificial drying methods are more complex and costly; however, they offer more control over the drying process. Both natural and artificial dryers can be built with widely available, inexpensive materials, helping to mitigate against cost. This chapter also covers options for assessing seed moisture content, which can help determine if seeds are dry enough to store.

Principles for Safe Drying

Seeds should be dried in ways that do not damage them or reduce their capacity to grow well when planted. Drying injury is a consequence of drying at too high of a temperature and/or too fast. Heat injury causes seed embryo mortality and cracking, shrinking, or splitting of seed coats.

A safe temperature for drying depends on the crop, how much moisture is in the seed, and the oil content of the seed. Seeds that have a high initial moisture content and that are high in oil are the most susceptible to heat damage. A drying temperature that does not exceed 40°C is safe for most seeds; however, a maximum temperature closer to 35°C is preferable for seeds with initial moisture levels (before drying) of over 15% (Hill and Johnstone, 1985).

Drying rate is linked to air temperature, air circulation, and humidity. Increasing temperature and air movement around the seeds will help seeds dry faster. High humidity will slow down the drying process. Aim for a balance between drying fast enough to prevent mold, and for efficient use of time, but slow enough to avoid damage to or incomplete drying of the seeds. Rapid drying is most likely to damage seeds when combined with high heat. Comparisons of drying temperatures of 35, 42, and 50°C found that heat damage and rapid drying (exceeding 2% moisture loss per hour) of shelled maize seed only occurred at 50°C (Herter and Burris, 1989).

Seed drying should also be safe for people. Consider the dangers of fire and electric shock in designing, building, and operating any drying technology that uses electricity.

Natural Drying Options

Natural drying is perhaps the simplest and, depending on materials used, can be the most inexpensive approach to seed drying. Control over temperature and humidity, however, is limited. Monitor temperature and humidity. During sunny, hot weather, move or cover seeds to protect them from overheating. Cool, cloudy weather prolongs outdoor drying time. If multiple days are required for complete drying, consider bringing seeds indoors at night to keep them from rehydrating. The risk of rehydration at night increases under high humidity and low dewpoint (indicating a high chance of dew).

Air drying on tarps, racks, and screens

Drying seeds on the ground with a tarp, screen, or other containers.

Air drying removes field moisture from harvested seed material such as panicles (a type of flower cluster or grain head) and pods. Spreading seed material out on tarps or racks is a good way to prepare them for processing and additional drying. A pre-drying step, for example, makes it easier to remove sorghum (Sorghum bicolor) seeds from the panicles. The processed seeds can then be placed in a seed dryer for further drying if needed.

If drying on the ground, keep seeds free of dust and particles of soil or gravel by placing the seed material on a tarp. Beware of high temperatures if drying in direct sunlight and move the seeds to a shaded location if necessary. A shelter or covering may be needed to protect seeds from birds or animals.

Above-ground drying with an A-frame rack.

Above-ground drying improves air circulation for more efficient drying. Options for this include the A-frame drying rack and a drying rack suspended near the underside of a metal roof.

A rack suspended near a metal roof.

Solar Drying Options

There are a variety of approaches to utilizing sunlight for drying seeds. Perhaps the simplest approach is to dry seeds under plastic as shown in the figure right. Using opaque instead of clear plastic will minimize sun damage by blocking some of the light coming in. Open sides will increase air circulation and reduce humidity.

Air-drying of seeds under opaque plastic.

With a chimney-type dryer the seeds are dried in an enclosed chamber, protected from pests and any adverse effects of direct solar radiation. A heat collector exposed to the sun generates heat that rises upwards into the drying chamber and exits through holes at the top. Spread seeds thinly on drying trays inside the chamber, which will help with uniform exposure of seeds to warm air in the drying chamber.

Chimney solar drier for drying plant biomass or seeds.

Artificial Drying Options

Seed drying cabinet

Adapted from: *ECHO Asia Note* 31 (Bicksler, 2017)

An effective seed dryer will include:

- a heating source
- a way to moderate the heating source
- a container or box for maintaining the heat
- screens on which to place seeds
- a fan to circulate air (optional)
- a vent for releasing moist air (optional)

Heated air from a heating element flows upward due to convection or is circulated around an enclosed space using a fan; this heating element is set to a specific temperature so as not to damage seeds. As the warm air moves through the chamber, seeds release moisture because warm air holds more moisture than cold air. If the chamber is very air-tight, your seed dryer will be more efficient with a vent to give off the warm, moisture-laden air since seed moisture will equilibrate to the ambient relative humidity.

Drying times and ability to reduce seed moisture content will vary depending on the efficiency of your dryer, the heating source, the quantity of seed being dried, the types of seed being dried, and the ambient conditions. It will be easier to dry seeds under dry ambient conditions. If the dryer is surrounded by humid air, it may be necessary to adjust the thermostat for increased heat (being careful not to overheat the seeds) and allow humid air to escape (which can be done via an exhaust port). Learning how to dry seeds properly requires trial and error and is both an art and a science.

Seed cabinet dryer equipped with drawers (left), hardware for removable screen shelves (middle), light bulbs as the source of heat (middle), thermostat for temperature control (right), and DC computer fans for air circulation (right).

With time and experience, it is possible to develop a keen finesse for drying seeds and approximating seed moisture content.

A high-quality and highly effective seed drying cabinet can be built from local materials with very low financial and labor inputs. The table below explains considerations for the various components of a cabinet dryer.

Component	Concepts and Considerations
A heating source	A heat source can be a heating element, light bulbs, or heat pads. Incandescent light bulbs give off the most heat, but fluorescent bulbs can be used. Inexpensive heat pads are available in various sizes and voltages. Electricity will be needed to power the heat source. Match power supplies to the power requirement of your heat source.
A way to moderate the heat source	The heat source can be controlled with either a thermostat or a simple electric timer. The thermostat is the preferred method, as it can accurately regulate temperatures. For smaller designs, 5- or 12-volt heat pads can be controlled with a 5- or 12-volt temperature controller such as the W1209 controller. A timer may be the cheapest option but is harder to use. The principle behind the use of a timer is to turn the heat source off and on at a frequency that achieves the desired temperature. Use of a timer requires time and effort to measure temperature inside the dryer, adjust the on/off function of the timer accordingly (to achieve a desired heating and cooling cycle), and to monitor changing ambient conditions and make adjustments going forward.
A drying chamber	This space will accommodate the seeds being dried and contain the heat that is essential for drying them. Containment of heat within this space is important for efficient use of the heat source and ensures that seed moisture does not re-equilibrate to ambient air moisture.
A way to contain the seeds	Screens work well to hold the seeds inside the drying space. Any material, such as screen material for windows or doors, that allows for airflow can work. The screens should be arranged to allow for maximum airflow through the chamber. They can be stacked in layers like shelves. Spread seeds no more than one layer thick on the screen material, to allow for good air circulation and to maximize drying.
A way to circulate air	A small fan that circulates the warm air can help to speed up drying times, maximize efficiency, and ensure more uniform drying. You could use an AC (alternating current) fan connected directly to a mains power source, or a DC (direct current) computer fan connected to a DC converter. Although helpful, a fan is not absolutely necessary if the heat source is below the screens and convection can carry the warm air up through the layers of screen shelves.

Component	Concepts and Considerations
A way to exhaust humid air and bring in fresh air	As warm air picks up moisture from the seeds, it loses its ability to hold more moisture, increasing the moisture within the drying chamber. A vent at the top of the chamber, that can be opened and closed, will allow warm, humid air to easily escape, which can be helpful if your seeds have a very high moisture content. An intake port will replenish humid air with drier air. Locating the intake port near the heat source will allow incoming drier air to be heated as soon as possible after entering the dryer. The vent and intake can consist of small holes or short pieces of PVC pipe with valves. Open the valves all the way to allow for maximum venting and cooling. Partially open or close the valves to increase air recirculation and heat within the dryer. You may not need a dedicated intake or vent if your dryer is not airtight.

Simple drying boxes

For small quantities of seed a drying box may suffice. One approach is to generate heat at the top of the box and force heated air downwards, via a blower fan, into the drying space.

Another approach is to heat and circulate incoming air at the bottom of the box and vent warm humid air at the top. The box can be something as simple as a Styrofoam cooler, with a temperature controller and DC fans as described on ECHO's GitHub page on microcontroller devices: http://edn.link/fp6qfe.

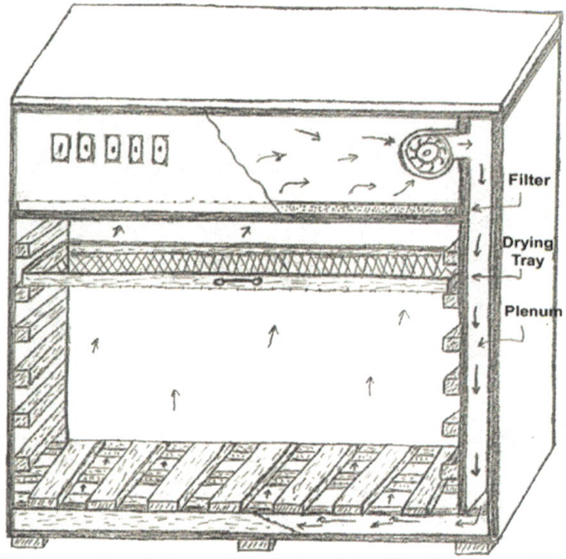

Diagram of seed drying box designed by Charlie Forst, with bulbs (heat source) and a fan in a top chamber. Air directed into a plenum wall (gap) on the right-hand side of the box exits at the bottom and circulates in a drying space that accommodates screen shelves. A filter keeps debris from contacting the heat source.

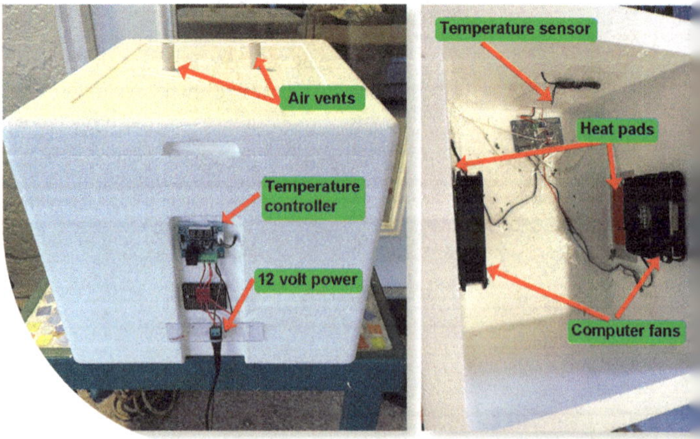

A temperature-controlled drying box equipped with DC computer fans for air circulation.

Desiccant drying

Desiccants are moisture absorbing materials that lower the humidity of the surrounding air. Desiccant options include rice, silica gel, and Zeolite Drying Beads®. Desiccant drying is best suited for small quantities of high-value seed. It is a good finishing option to bring seeds to a target moisture level after an initial air-drying step. The main requirements for desiccant drying are a sealed container and desiccant material that is dry and thus able to absorb moisture. Optional additions could include fans, for air circulation, or a mechanism to rotate the container periodically for more uniform drying. Place the desiccant in a sealed container with the seeds to be dried. If possible, monitor humidity in the container to ensure that the desiccant is keeping humidity low. Chapter 11 will contain more information on desiccants and technologies for monitoring humidity.

How to Determine if Seeds are Dry Enough?

Aim for an acceptable seed moisture level

It can be helpful to distinguish between ideal and acceptable seed moisture. In general, orthodox seeds should ideally be dried to a moisture content of between 3% and 7% for long-term storage (Ghimire and Joshi, 2019). Recommended moisture contents are lower for oil than non-oily seeds. Water has an affinity for protein and starch but does not mix with oil (McDonald 2007). Thus, at any given level of humidity, the moisture

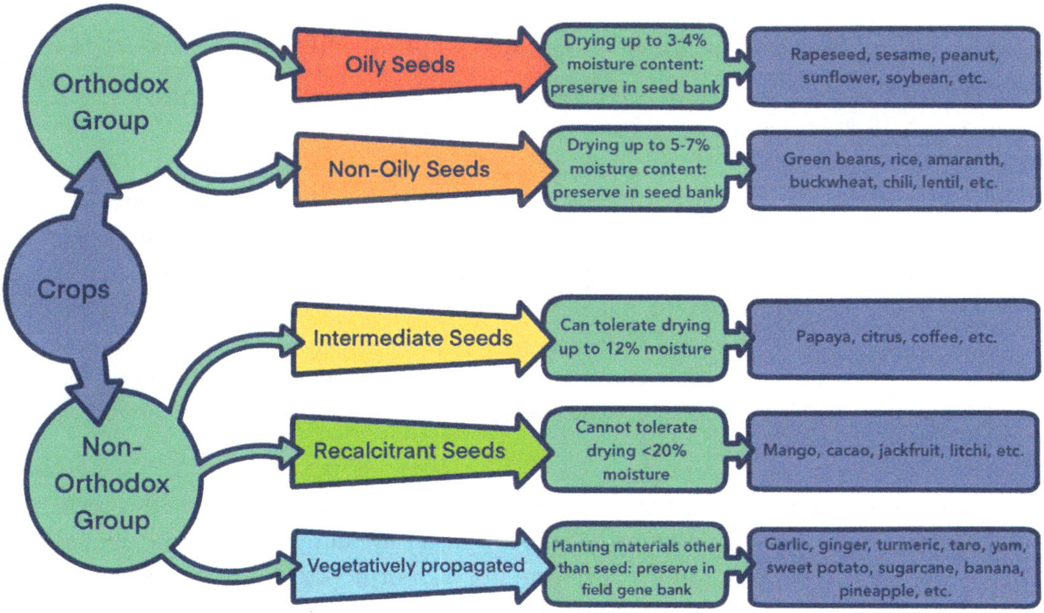

Source: [Ghimire and Joshi [http://edn.link/a6z6cz]](http://edn.link/a6z6cz), 2019. Image adapted by Arielle Price

content of seeds high in oil will stabilize at a lower level than that of non-oily seeds high in protein or starch. Examples of seeds high in oil, protein, and starch are shown in the table below.

High starch (50%)[1]	High oil (18-50%)[1]	High protein (18-50%)[1]
Amaranthus spp. (amaranth)[2]	*Arachis hypogaea* (peanut)[1]	*Glycine max* (soybean)[6]
Avena sativa (oats)[1]	*Glycine max* (soybean)[1]	*Lespedeza* spp. (lespedeza)[1]
Chenopodium quinoa (quinoa)[3]	*Gossypium* spp. (cotton)[1]	*Medicago sativa* (alfalfa)[1]
Daucus carota (carrot)[1]	*Helianthus annuus* (sunflower) [1]	*Moringa oleifera* (moringa)[1]
Oryza sativa (rice)[1]	*Linum usitatissimum* (flax)[1]	*Phaseolus vulgaris* (common bean)[1]
Sorghum bicolor (sorghum)[1]	*Moringa oleifera* (moringa)[4]	*Vicia faba* (faba bean)[5]
Triticum aestivum (wheat)[1]	*Ricinus communis* (castorbean)[1]	*Vicia villosa* (hairy vetch)[1]
Zea mays (maize)[1]	*Sesamum indicum* (sesame)[5]	*Vigna unguiculata* (cowpea)[1]

Information gathered from: [1] Potts (1972), [2]Malik et al. (2023), [3]Guantian and Zhu (2018), [4]Anwar et al. (2005), [5]El-Khier et al. (2008), and [6]Martineau-Côté et al. (2022).

Ideal moisture levels for long-term preservation of orthodox seeds may not be realistic or practical for seed banks with limited options for controlling temperature and humidity. Moreover, while long-term preservation is important for gene-banks, CSBs may not need to store seeds for much more than a year, in which case drying to a seed moisture content of 10 to 12% is acceptable for most orthodox seeds (Chala and Bekana, 2017). Fungal growth becomes problematic at seed moisture levels above 13% (McCormack, 2010). The life span of stored seeds will also depend on temperature, as will be discussed in Chapter 11.

The figure on the previous page also addresses moisture levels for non-orthodox seeds. Recalcitrant seeds have little or no tolerance for drying and should be distributed/planted soon after harvest. On the spectrum between orthodox and recalcitrant seeds are those with intermediate tolerance to drying. It may be possible to store seeds in the intermediate category, but they generally have a shorter storage life and are less tolerant of freezing temperatures than orthodox seeds (Trusiak et al., 2023).

Measure Seed Moisture

Adapted from: *ECHO Development Note* 136 (Reader and Tim Motis, 2017)

To know if seeds are dry enough to store, it is necessary to assess seed moisture. Here we present options for doing so. We will start with methods that require instrumentation or an oven and give precise measurements of seed moisture. We will also present techniques that provide a rough indication of seed moisture, requiring no special equipment.

Moisture Meters

Seed moisture meters can be helpful for organizations or groups that need frequent and/or precise seed moisture readings. Some meters work by heating a small sample of seeds (a range of appropriate grams of seed material [typically crushed in a mortar and pestle]) and displaying the moisture content after the weight stabilizes. Other meters make use of the fact that electrical conductance varies with seed moisture. While effective, moisture meters may be inappropriate and/or unaffordable for most farmers.

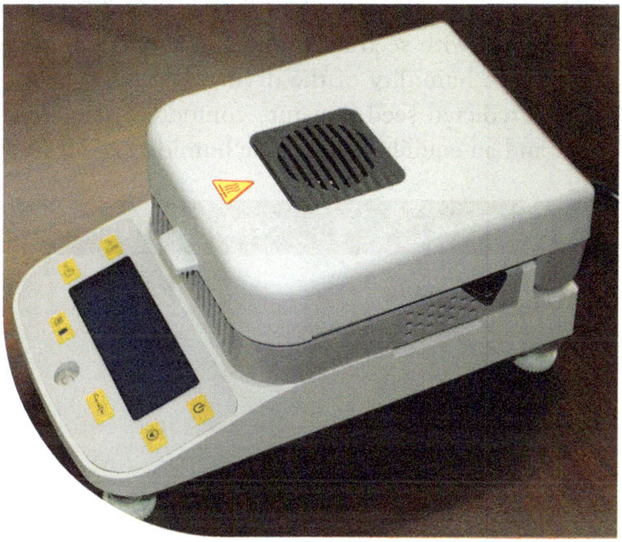

Moisture analyzer that uses heat to achieve a constant dry weight.

Oven Drying

Seed moisture content can also be determined with an oven as described below, based on guidelines established by the International Seed Testing Association. Take a random sample of your seeds and obtain the fresh weight. Then, place the fresh seeds in the oven, wait for the temperature to reach 103°C (oily seeds) or 130°C (non-oily seeds), and maintain that temperature for an appropriate amount of time: 17 hours for oily seeds and 1 to 4 hours for non-oily seeds (4 hours for maize, 2 hours for other cereal crops, or 1 hour for other crops with non-oily seeds). Next, remove the sample from the oven and allow the seeds to cool. Finally, weigh the seeds to obtain the dry weight. Use the following formula to calculate seed moisture content as a percentage of the initial weight before drying (ISTA, 2005; Rao et al., 2006).

$$\% \text{ moisture content} = \frac{\text{weight of fresh seeds - weight of dry seeds}}{\text{weight of fresh seeds}}$$

Drawbacks of this method are that the seed sample is destroyed, and that many small-scale farmers may not have access to an oven. Some simpler tests do not require instrumentation or costly equipment.

Measure the equilibrium relative humidity

Dry seeds will absorb moisture if they are surrounded by humid air, thus lowering relative humidity of air around the seeds. Conversely, moist seeds surrounded by dry air will release moisture to the air, thus raising relative humidity of air around the seeds. In a closed space, the relative humidity of air in the drying space will reach equilibrium with seed moisture. The **equilibrium relative humidity** level, therefore, is the relative humidity of the drying space at which seeds stop releasing or absorbing water. Predicted seed moisture content of selected crops at a drying temperature of 30°C and an equilibrium relative humidity of 50% are listed in the table below.

Species	Seed oil content (%)	Predicted seed moisture (%)
Amaranthus caudatus	6.5	10.3
Arachis hypogaea	46.8	6.1
Brassica oleracea var. acephala (kale)	25.9	8.3
Glycine max	20.0	8.9
Helianthus annuus	39.9	6.8
Lycopersicon esculentum	22.5	8.6
Moringa oleifera	37.5	7.1
Sorghum bicolor	5.0	10.4
Tamarindus indica	7.0	10.2

Data extracted from the Seed Information Database (SID) hosted by the Society for Ecological Restoration (SER) through the International Network for Seed-based Restoration (INSR); SER, INSR, RBGK, 2023.

Species	Seed oil content (%)	Predicted seed moisture (%)
Vicia faba	3.9	10.5
Zea mays	4.9	10.4

Data extracted from the Seed Information Database (SID) hosted by the Society for Ecological Restoration (SER) through the International Network for Seed-based Restoration (INSR); SER, INSR, RBGK, 2023.

You can measure relative humidity with inexpensive humidity indicator strips or loggers (hygrometers), as discussed further in chapter 11. Equilibrium relative humidity will be the point at which the relative humidity stabilizes (stops changing). To measure equilibrium relative humidity in about 30 minutes, place a sample of the seeds being dried, and a humidity logger or indicator strip in a small, sealed container (Sutcliffe and Adams, 2014). The container should be nearly full of seeds; if using a logger make sure the humidity sensor is not touching the seeds. The test will be most accurate if done at the same temperature as the seeds; if drying seeds in a drying cabinet, for example, conduct the test with the sample container in the drying cabinet.

Bite and Bend Tests

One common technique for determining if bean seeds are dry enough to store is to bite or press a fingernail into a seed. If no mark is left on the seed coat, the seed is most likely dry enough for storage. Seeds of cucurbit species can be tested by bending them; sufficiently dry seeds should not bend easily.

Salt-Jar Test

The salt-jar test is another simple option. This test is based on the fact that salt clumps at relative humidity levels of 70-75% (Sutcliffe and Adams, 2014). Seed moisture content is affected by relative humidity. At 70-75% relative humidity, the moisture content of maize seed stabilizes close to 15% (Mrema, 2011).

Step 1: Ensure that the inside of a clean jar is dry. If you think it may not be completely dry, remove the lid and place the empty jar and lid in a dry area for a time (30 minutes to an hour). Options for a dry area could range from outdoor air during a time of day when humidity is low, or in an oven at the lowest heat setting. Now add 5 ml (1 teaspoon) of salt to the jar, put the lid on tightly, and roll the jar with the salt inside. If the jar is dry, salt will not stick to the sides.

Step 2: Remove the lid and pour maize seeds and salt into the jar, leaving enough empty space to be able to shake the seeds and salt. The amount of salt added with the seeds is not critical. Sutcliffe and Adams (2014) suggest a 1:1 ratio of salt and seeds, with the seeds and salt filling half the volume of the jar.

Step 3: Put the lid on the jar and shake the jar, rolling it gently for 2 to 3 minutes. If the salt does not form clumps or stick to the sides of the jar, the moisture content of the maize seed is probably less than 15%.

Salt jars with subsamples of maize seed; salt at bottom of jar indicates dry seed.

Salt sticking to glass jar indicates wet seed.

Chapter 11. Low-Cost Storage Technologies

Proper Storage Conditions

Adapted from: *ECHO Best Practice Note 5* (ECHO staff, 2016)

Storage life of orthodox seeds is favored by low seed moisture and cool temperature. Seed moisture is kept low by minimizing humidity in storage rooms and/or containers. In managing humidity and temperature of the storage environment, ECHO seed banks follow the "100 rule" below:

Percent relative humidity + air tempearture in degrees F = 100 or less

(Harrington, 1972; McDonald and Copeland, 1997)

For example, if the temperature in a seed storage room or container is 80°F, the "100-rule" can be satisfied by keeping relative humidity to 20% or less. Between temperature and humidity, humidity is the most important factor to control. This is because, as relative humidity reaches 70%, seed moisture content approaches 13%, at which point fungi start to become a problem in storage (Justice and Bass, 1978; McCormack, 2004). For most seeds, multi-year storage is possible at less than 50% humidity, especially if temperature is also kept low (close to 50 °F; McCormack, 2004). A humidity level of 50% is well below the 70% threshold for mold growth.

Refrigerators, walk-in coolers, and dehumidifiers offer a high degree of control over temperature and/or humidity. These options, discussed in the subsequent chapter on infrastructure, require reliable electricity and sufficient monetary and material resources. The focus of this chapter is on low-cost technologies for keeping seeds dry and cool.

Keeping Seeds Dry

Containers

Container options include paper envelopes, plastic bags, laminate foil packets, cans, jars, and buckets. Weight, shape, and size are several considerations in selecting containers to store seeds in. Envelopes, bags, and packets are lightweight and flat, making them easier to store and transport than cans, jars, and buckets. Buckets are a good option for storing bulk quantities of seeds.

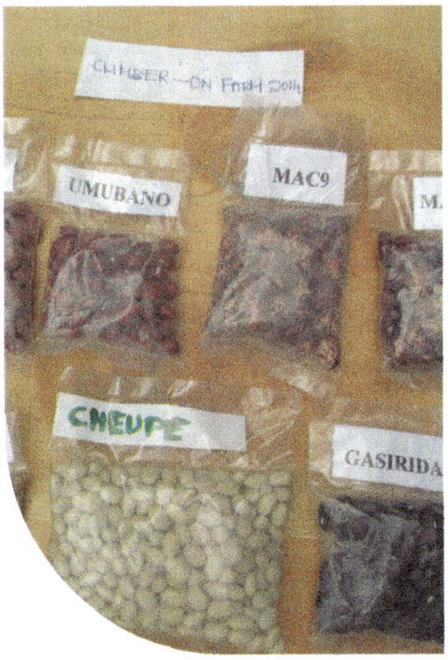

Other important factors in selecting containers are the humidity of the surrounding air and how well they keep water vapor from entering the container. Paper is porous, allowing water vapor to easily enter the envelope. Thus, paper envelopes are acceptable for a dry storage room or container but a poor choice for a humid environment. Plastic bags vary in thickness and density. Thin, low-density polyethylene not only allows slow passage of water vapor over time, it also is more easily punctured and torn. If possible, select high density polyethylene bags with a thickness of 4 mil (100 microns) or more (Wilnan, 1987). Consider double bagging (placing one bag inside another) if thin plastic bags are the only container available. Of the flat container options, laminate aluminum foil or mylar packets are the most resistant to water vapor. They are available in various sizes, thicknesses, and sealing mechanisms (e.g., zip-lock and/or heat seal). Cans, jars, and buckets exclude humidity very well if they have lids that provide an airtight seal.

A container with an airtight seal is said to be **hermetic** or **hermetically sealed.** Hermetic sealing prevents the movement of both water vapor and oxygen from outside air into the container. Hermetic storage of orthodox seeds is only effective if the seeds are dry initially. Otherwise, water will be trapped in the container, creating moist conditions that favor mold. Hermetic storage can be further optimized with vacuum-sealing and desiccants.

Vacuum-sealing

Vacuum-sealing makes use of airtight/hermetic containers to not only exclude ambient air but to also remove air—along with water vapor and oxygen the air contains—from the container. Vacuum sealing can be done with simple modifications to low-cost devices including bicycle pumps, automobile brake bleeder pumps, and syringes. See *ECHO Technical Note* 93 (http://edn.link/tn93; Motis, 2019) for instructions on how modify and use these items for drawing vacuum on a container of seeds.

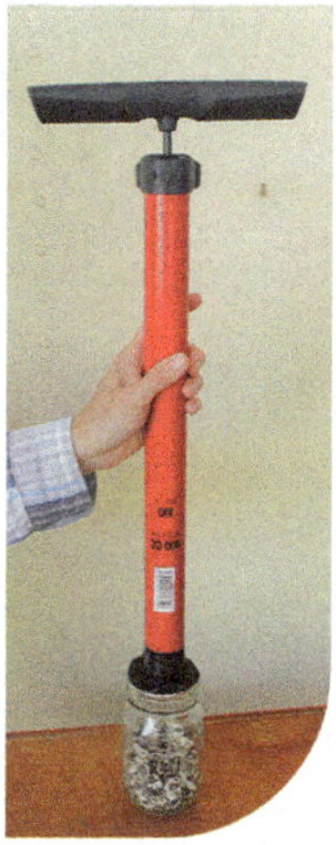

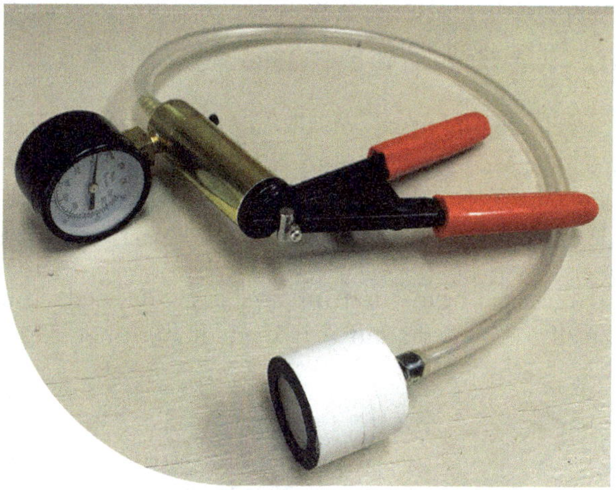

Desiccant Options and Methods

Controlling the humidity in an entire room requires electricity to run a dehumidifier. Desiccants in sealed containers offer a lower-cost alternative that is most practical for storing small quantities of high-value seeds. Desiccants reduce the humidity of the surrounding air by taking in moisture. Materials that ECHO is most familiar with are rice, silica gel, and Zeolite Drying Beads®.

Rice

Uncooked white or brown rice is widely available and inexpensive. Rice kernels gain or lose moisture depending on how much moisture is in the kernels and the surrounding air (Kunze, 1977). We recommend using de-hulled rice kernels, as the hull may interfere with the ability of the inner kernel to gain moisture.

Rice with a high moisture content may take in little or no moisture from the air. Thus, it is important to remove existing moisture in the rice before use (*BPN* 5). Spread rice out on a tray placed in the middle rack of a household oven and bake the rice for an hour at 150 °C (ECHO Staff, 2016). Alternatively, heat the rice in a pan over an open flame (Morgan, 2013).

Once the rice kernels are lightly browned or stop losing weight, place them in the container(s) to be used for storing seeds. Drying to a constant weight is the best way to be sure that the rice is completely dry. To do this, take the rice out of the oven every 15 minutes or so, weigh it, put it back in the oven and repeat the process until there is no more weight loss. Allow the container to cool for about 20 minutes before placing seeds into it. A good starting point, in terms of volume of rice, is to fill a container 25% of the way full with heated rice. To maintain the effectiveness of the rice, keep the container closed/sealed; also close it again right away after removing seeds.

Rice as a desiccant placed in the bottom of a sealed container with envelopes of seed placed over the rice. Note the wireless hygrometer used to monitor humidity.

Silica gel

Silica gel consists of silicon dioxide (silica), a compound found in sand, processed into beads with tiny pores that hold water. The small beads are often packaged in packets made of cloth or paper and used to keep electronics dry.

Keep silica gel beads from mixing with the seeds. Silica gel itself is not toxic, but it is inedible and may contain indicator beads that are not safe to eat (Warneke, 2020).

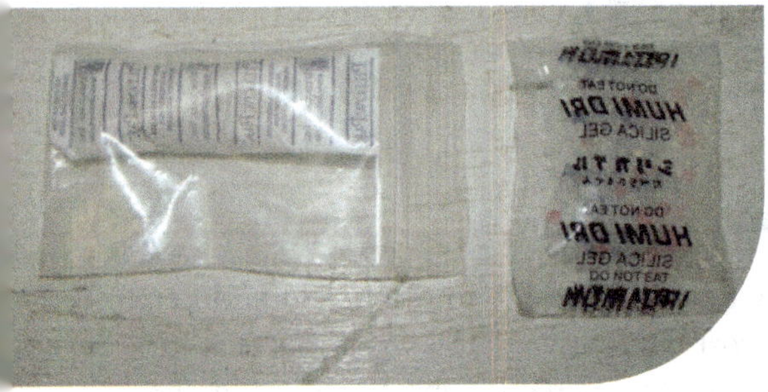

Silica gel packets with (left) and without (right) indicator crystals

For keeping seeds dry in storage, start with a ratio of 1 part silica gel to 10 parts seeds (e.g., 1 kg silica gel with 10 kg of seeds as suggested by Copeland and McDonald, 2001) and adjust as needed. If techniques used to monitor moisture in the container indicate that the desiccant is not keeping seeds dry, it will be necessary to increase the desiccant quantity and/or replenish the desiccant.

Silica gel beads that have become saturated with moisture can be regenerated many times with heat. To do so while maintaining the color-changing capacity of indicator beads, heat silica gel in an oven at a temperature between 105 and 120°C (Conservation By Design Limited, 2012; Warneke, 2020).

Zeolite Drying Beads®

Using zeolite, Rhino Research in Thailand developed a desiccant product called Drying Beads®. An advantage of zeolite beads over silica gel is that they are better able to hold moisture at low humidity (Hay et al., 2012).

Zeolite beads can reduce seed moisture content to less than 5%. Too much zeolite in a container of seeds can injure the seeds due to over-drying. For example, a ratio (by weight) of 2 parts seed to 1 part beads works well for storing sorghum (*Sorghum bicolor*) seed but reduces germination of okra (*Abelmoschus esculentus*) and velvet bean (*Mucuna pruriens*) seeds (Trail et al., 2022).

The quantity of beads needed to achieve good results is influenced by factors such as crop, humidity, initial seed moisture, and the quantity of seeds to be dried. See the website, dryingbeads.org [http://edn.link/zkj222], to access a tool for calculating the amount of beads to use.

As with silica gel, zeolite beads can be reused. To do this, heat the beads in an oven for 2 hours at a temperature of 250°C (FFIL, undated).

Alternative desiccants

Other examples of desiccants include calcium oxide and charcoal. Calcium oxide is inexpensive and often sold as burnt lime. ECHO research has shown that it is capable of stabilizing seed moisture content over a one-year period (Trail et al. 2022). To avoid inhaling fine dust, work slowly when transferring calcium oxide powder to seed containers. Calcium oxide cannot easily be reused like silica gel or zeolite beads; this is because it changes to a different substance (calcium hydroxide) as it reacts with water.

Desiccants like silica gel will most likely be dry enough to use when initially purchased; however, like rice, charcoal needs to be dried before using it to dry seeds. Charcoal can be dried to a low moisture content by placing it under direct sun for 5 hours. 200 g of dried charcoal works well for storing 40 g of sesame seed over a three-month period (Oyekale et al., 2014). To maximize the surface area, and thereby improve the drying capacity, break the charcoal into small pieces or grind/pound it to form a powder.

Two ways to use desiccants

There are two main approaches to using desiccants in seed banks:

- First approach- to dry seeds down initially, in preparation for storage
- Second approach- to keep seeds dry in storage

The first approach works where seeds are 1) kept in a dry, climate-controlled storage area or 2) kept in sealed containers such as jars (instead of thin plastic bags) that are thick enough to effectively exclude humid air and will not be opened for a long period of time. Place the desiccant with seeds in an airtight container and then remove the desiccant once the seeds are dry.

Use the second approach where 1) ambient humidity is high for at least part of the year (e.g., rainy season) and 2) where electricity and equipment for dehumidifying a room is unreliable or too expensive. In this situation, store seeds with the desiccant to keep them dry in storage.

How to assess desiccant performance

As containers are opened and closed to access seeds, humid air may be reintroduced into containers depending on ambient conditions. Over time, a desiccant will become saturated and unable to take in more moisture. Below are three ways to determine whether or not a desiccant needs to be replaced or regenerated.

- Monitor humidity of a seed storage container with a hygrometer. Replace or regenerate desiccants when you see a trend of increasing humidity over time.
- Place one or more packets of silica gel with indicator beads in the seed container. A color change will serve as an indicator that the silica gel (and possibly other desiccants) should be replenished.

The above two methods do not necessarily need to be implemented in every container. Rather, use one or more methods to monitor seeds in one or two containers that serve as a sample of similar containers. For example, for 10 same-size containers of cowpea seed filled on the same date, a seed bank manager may decide to only monitor one of the containers. If the 10 containers are being opened and closed at different times to access seeds, then it may be best to employ one of the above techniques in more than one container.

Management of Stored Seed Pests

Various beetles, weevils, and grain borers can damage stored seeds. The nature of the damage they cause varies with species. Bruchid beetles, for example, damage the endosperm, creating holes in the seed coat. Others feed on the embryo, rendering seeds unable to germinate. As insects feed on stored seeds, they also generate heat, moisture, and waste, all of which contributes to spoilage.

Cowpea (*Vigna unguiculata*) seed damage...

caused by bruchid beetles (*Callosobruchus* sp.)

Below are practices that prevent or reduce seed losses to insect pests:

- **Manage pests in the field.** This will minimize the number of insect eggs deposited on seeds in the field and, thus, the number of eggs that can hatch in stored seeds.
- **Store seeds cool and dry.** Heat and humidity favor insect feeding activity and reproduction. Proper storage conditions, as described earlier in this chapter, are therefore not conducive to rapid population growth of storage insect pests.
- **Deprive insects of oxygen.** Air contains 20-21% oxygen. Seed storage insect pests may not need much oxygen to survive but a low-oxygen environment will slow feeding and reproduction. Insects such as cowpea bruchids stop feeding when oxygen is reduced to 2% (Ahn et al., 2013). To achieve a low-oxygen environment, keep dry seeds in sealed, airtight containers. Any insects present will naturally respire, using up oxygen and releasing carbon dioxide. A combination of low oxygen and high caron dioxide does not harm dry seeds—since they respire at such a low rate—but insects will become inactive and eventually die. Optimize the effectiveness of airtight sealing by filling the containers full (to displace oxygen). For even greater effectiveness, remove air from sealed containers via vacuum sealing (Lawrence et al. 2017).
- **Use local treatments.** A number of options have been assessed, including vegetable oil, pulverized bamboo charcoal, galangal powder, powdered detergent, bleach, and carbaryl. Bleach and oil adversely affected seed germination, but the remaining treatment options maintained high germination levels over a one-year storage period (Lawrence et al. 2017).

Monitoring Storage Conditions

Monitor storage conditions to ensure that temperature and humidity are within acceptable ranges. Below are four technologies for doing this:

Humidity indicator cards are an inexpensive option and available from various suppliers. Each card will typically have multiple circles, with each circle representing a given level of humidity. Within each circle, a change in color from blue (dry) to pink (humid) means that the level of humidity indicated for that circle has been reached.

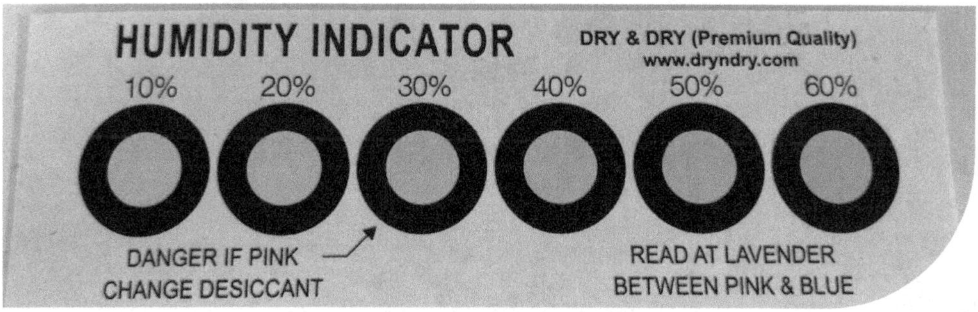

A humidity indicator strip.

Wireless hygrometers are another low-cost option and widely available from online vendors. A hygrometer will display both humidity and temperature and is typically powered by a battery. The most inexpensive units simply display temperature and humidity on a small screen. Simply place the hygrometer in the space you want to monitor, and view/record readings as needed. Some hygrometers come in two parts, a transmitter and a receiver, allowing you to place the transmitter in a container or room you want to monitor and the receiver in a convenient location for viewing temperature and humidity readings; this setup allows for tracking of storage conditions inside a sealed container without having to open the container to view the readings.

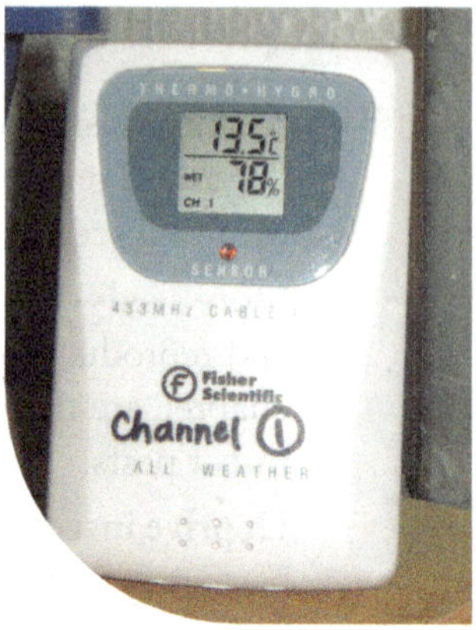

A wireless hygrometer.

A thermometer can be placed in your seed storage room to indicate temperature.

Salt jar testing, described in Chapter 10, will indicate humidity. Place a sample of seeds in a jar along with 5 ml of salt and screw on the lid. Gently shake or roll the jar. If the salt sticks to the sides of the jar your humidity levels are probably too high. Humidity is likely to be at an acceptable level if the salt does not stick to the sides of the jar.

Chapter 12. Infrastructure Options for Seed Storage

Adapted from: *ECHO Asia Note* 27 (Price, 2016) and 38 (Trail et al., 2019).

Examples of seed bank structures made with local (left) or re-purposed (right) materials.

Infrastructure Considerations for Seed Storage

Given the proper facilities necessary to store seeds long-term, whereby low temperature and low humidity are kept stable over time, it is very possible to store most orthodox seeds for several years at a time in the tropics (Harrington, 1972). Unfortunately, implementation and maintenance of the proper facilities can be very costly and many existing seed bank and gene bank facility examples do not satisfy the needs of many smaller organizations or communities.

Thankfully, many diverse options currently exist, with varying levels of investment for a wide range of facilities, from expensive, high-tech facilities down to low-cost, low-maintenance models. At ECHO, we operate our own range of seed storage facilities at our various seed banks around the world, from a high tech, walk-in climate-controlled cold storage room, to a retrofitted refrigerated shipping container, to a low-cost, foam-insulated cold room cooled with a standard split-unit air conditioning system.

While each of these systems has proven to be effective in storing seeds for our needs over time (Motis, 2016), even these lower cost options do not adequately address the very real question of how seed storage facilities may be replicated at a farm or local community level. This chapter features a range of infrastructure options including some that do not require a constant supply of electricity.

A few important factors to consider with respect to infrastructure are:

- **Scale and purpose:** Will the seed bank serve a small community or an entire district or region? What crops and quantity of seed of those crops will the seed bank carry and offer to clients? Will year-to-year storage be sufficient, or will the seed bank need to store seeds for multiple years?
- **Material and financial resources:** Seed banks constructed with readily available, low-cost materials are the easiest to replicate.

Over the years, ECHO has scaled its various seed banks around the world to the specific needs of the individual seed bank. Each has its appropriate scale and serves different purposes, from distributing thousands of seed packets a year, to serving the needs of a single community or region. The table below summarizes the costs of implementing these different options. Note that these prices may vary considerably depending on your location and access to materials and/or labor, and existing structures being retrofitted.

The table below summarizes the economics of various Seed Bank Cold Room and Seed Storage options used at ECHO's various sites around the world are.

	Structure	Location	Details	Size (m)	Avg. Temp (°C)	Cost (USD
Long-term storage options	Cold room	Florida, USA	High-tech, climate-controlled, walk-in	6 x 3 x 2.5	5	35,000
		Florida, USA	Retrofitted refrigerated shipping container, single split-unit air-conditioned (A/C) system with CoolBot sensor	2.5 x 3 x 2.5	10	10,000
		Chiang Mai, Thailand	Foam insulated, double split-unit A/C system with CoolBot sensor	8 x 5 x 2 .5	15	5500
		Mae Ai, Thailand	Foam insulated, single split-unit A/C system with CoolBot sensor	5 x 3 x 2	6	3250
Year-to-year storage options	Earthbag house	Mae Ai, Thailand	Clay and rice hull filled bags, stacked walls with thatched roof	4 (diameter) x 1.7 (height)	23	750
	Hillside bunker	Mae Ai, Thailand	Dug-out storage nook, in primarily clay-based hillside or sloping land	2 x 2 x 1.5	-	80
	Buried cistern	Mae Ai, Thailand	Large glazed ceramic cistern designed for water storage, buried up to the rim in the ground	-	23	20

Further details available in *ECHO Asia Note* 27 (Price, 2016).

Note that the higher-cost options listed above are designed more for multi-year storage of seeds (such as in a seed bank or gene bank setting), while the lower-cost options are designed more for community or farm-level settings where seeds only need to be stored year to year or two years at most.

Storage Facilities that Require Electricity

Refrigerated cold rooms

Refrigerated, walk-in coolers are capable of rapid cooling to desired temperatures, down to near freezing. They vary in size, functionality, and price. In addition to the electricity they consume, they require coolant and regular maintenance. Cooling technologies vary in their ability to control humidity; a dehumidifier may be needed.

Low-cost air-conditioned cold rooms

A standard window air conditioner will cool a room to about 15°C. For further cooling, to nearly 2°C, pair a window air conditioner with a CoolBot. Doing so can be done without tools or specialized training in electronics. The unit and installation instructions are available through the CoolBot website (http://edn.link/an33ya). A window air conditioner in combination with a CoolBot provides a low-cost approach to cooling a storage room.

A standard window air conditioner paired with a CoolBot.

A seed storage room in Asia cooled with a CoolBot system.

Humidity and condensation are major factors to consider in designing a cold room. The addition of a dehumidifier will help to lower humidity. A dehumidifier will typically feature a water collection pan and/or a drainage port. For ease of use, connect the drainage port to a garden hose or pipe that can be used to direct water outside. Otherwise, the collection pan will need to be emptied periodically to keep it from overflowing. See *ECHO Asia Note* 27 (http://edn.link/qgpe6j), for more design considerations including insulation and vapor barriers.

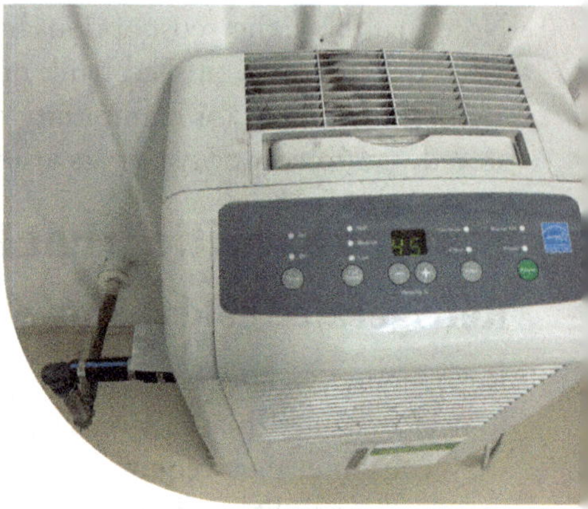

A dehumidifier with drain line.

A refrigerator

A household refrigerator will keep seeds cool but humidity can be high and can fluctuate (Croft et al., 2012). Seeds stored in a refrigerator should be kept in a sealed container. Vacuum-seal the container and/or add desiccant along with the seeds for greater effectiveness in keeping the seeds dry. When keeping seeds in a refrigerator do not store them in porous containers such as paper envelopes, as these will expose the seeds to humidity.

Natural Building Storage Options

It has been suggested by many that various natural building infrastructure such as earthbag and adobe houses may serve as ideal sites for storing seeds. Storing seeds underground has also been explored. Natural building structures can be constructed with local materials and do not require electricity. They stabilize and lower temperature but maybe humid (Trail et al., 2019). Keep seeds from absorbing moisture in natural

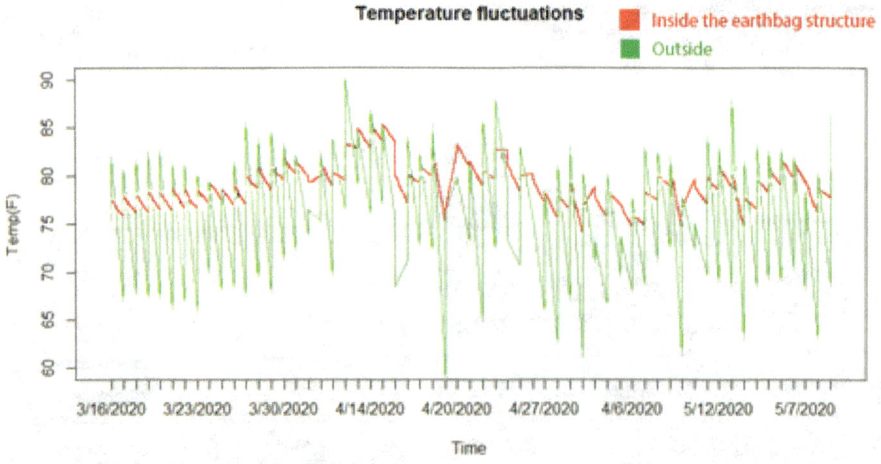

Temperature fluctuations inside (red line) and outside (green lines) in an earthbag structure in Thailand. Data collected by ECHO staff.

building structures by storing them in airtight containers in combination with vacuum-sealing and/or desiccants. Below are several options that ECHO has trialed over the years.

Earthbag seed banks

As explained in *ECHO Technical Note 96* (http://edn.link/tn96), heat absorbed by the thick walls during the day transfers to the interior space at night. This delayed transfer of heat is what moderates day- and night-time temperature (Kiefer et al., 2019). *ECHO Technical Note* 96 contains step-by-step instructions for building an earthbag structure.

Adobe structures, such as those pictured below, have similar advantages to earthbag structures.

Underground storage options

Another natural building approach to achieve cool, stable temperature is to keep seeds underground. Underground seed storage techniques include hillside bunkers and buried cisterns. Underground structures tend to be more humid than earthbag structures. With the use of vacuum-sealing, however, *Lablab purpureus* seeds kept in earthbag houses, hillside bunkers, and buried cisterns germinated at a level above 90% over the course of a year. Even with relative humidity over 90% in a buried cistern, the moisture content of vacuum-sealed seeds increased by less than 2% (Trail et al., 2019).

Underground storage techniques, showing a hillside bunker (left) and buried cistern (right).

Organization of Space

Whether a seed bank consists of a single or multiple rooms, organize the space in a way that facilitates ease of operation and health of the seeds. A few factors to consider are:

- Proximity of seed storage space in relation to space for seed processing: Separate these spaces if possible to minimize contact of stored seeds with contaminants (e.g., debris or disease) of seeds being processed and sorted.
- Temperature of space for storing and germination testing: Water can condense on seeds or containers when they are brought abruptly out of cool storage conditions into a warmer environment for germination testing or distribution. If you have the ability to cool the space for both storing and germination testing, cooling efficiency can be gained with those spaces adjacent to each other.

Mbegu

Part IV. Managing Seed Inventory

Chapter 13. Inventory Management

Record Keeping

Record keeping traces a seed variety from the time it is catalogued in the seed bank, through storage, testing, and replenishment at the seed bank, and finally, when and to whom it is given when seed from that variety leaves the seed bank. This process of keeping records ensures the availability and accuracy of information and labeling. Sufficient, accessible information improves the functioning and quality of the seed bank over time. For example, inventory of seed quantities, seed germination rates, and other information can be tracked and updated using clear and descriptive records.

Accession Numbering Systems

The first step of cataloguing a new batch of seed is to give it a unique identifying number called an **accession number.** This number is like a birth registration for that batch of seed and should not be reused if that seed is removed from the seed bank. It is helpful if the accession number includes the year it was acquired/grown, for quick reference. Individual seed banks create unique accession numbering systems for their own needs. In creating a system, make sure the procedure for numbering is consistent and that all workers are trained in the process so that numbers are not accidentally duplicated or mislabeled.

Here are two examples of accession numbering systems:

- FAO community seed banks (FAO, 2014):
    - 2 digits for year of harvest
    - 2 digits for the variety code
    - 2 digits for the farmer that produced the seed
    - 2 digits for the individual seed lot number
    - Example: 19221201
- ECHO seed banks:
    - 2 digits for year of acquisition
    - 3 digits for the number of accessions that year
    - 2 letters for the seed bank
    - 2 digits for the year the lot was acquired
    - 1 digit for the seed lot
    - 1 letter for the source of the lot
    - 1 digit for the location of the lot
    - Example: 12032TH-131E1

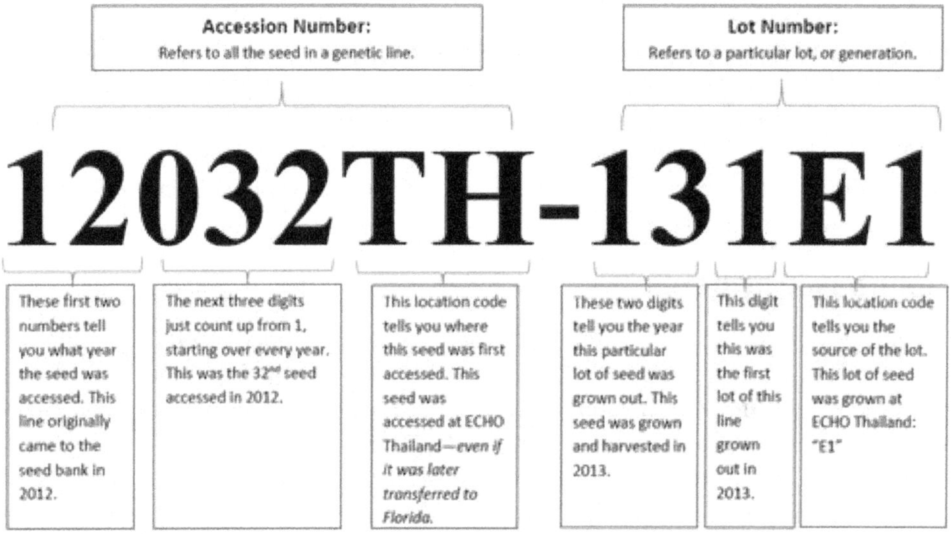

If a new batch of seed comes from an existing accession, such as when an accession is multiplied or acquired from the same source (same variety and supplier), the accession number stays the same but the lot number changes. **Lot numbers** indicate a new generation of a particular accession. When an accession is grown and harvested, the genetic line is the same, but the lot number differentiates the newly harvested seed from the original seed. As mentioned previously, make sure the procedure for numbering is consistent and that all workers are trained.

Accession Records

In addition to an accession number, record from whom and where the seeds were acquired, the date they were acquired, and the quantity. If seeds were purchased or grown out by a farmer, list the farmer's contact information and where they originally acquired that variety. Keep notes of unique characteristics of that variety, including traditional use, desirable traits, and nutrient content. Keeping this information is extremely helpful as people grow crops that are new to them or as people preserve their traditional crops. It is also important to weigh how much seed is coming in and going out, so every seed bank should have a scale (Vernooy et al., 2020).

To summarize, an **accession record** can include the Latin name of the crop; local name; variety name; accession and lot number; name of supplier and their contact information; location where seed was produced; amount of seed; provided germination rate; and typical characteristics of the variety – such as cultivation, resistance, nutritional value, and cultural uses. If money is exchanged for seeds or if the objective of the seed bank is to generate income, then it is important to keep financial records for income, expenses, and profits. Record all these things on a paper register or in an electronic database.

Write out, print, or copy a premade form for accession records when receiving new seeds to ensure that the necessary information is recorded. Accession records are kept on file in numerical order.

Some record keeping is required by law for seeds destined for sale, such as germination rates and date tested. Check local seed laws to comply with requirements. Some seeds are restricted by patents or research stations which developed them, in which case you must obtain and record permission to use them. A Memorandum of Understanding (MOU) or Material Transfer Agreement (MTA) outlines how a variety obtained from another organization can and cannot be used. Keep this type of document with the accession record for quick reference.

Paper based systems

Record systems range from simple to complex. Seed banks commonly start with the use of paper record keeping. Later, seed banks can combine paper records with electronic records for redundancy. Recording seed deposits and withdrawals are the main information to record. Some seed banks keep a simple logbook for the date of the deposit or withdrawal, the name of the farmer, the community, the name of the crop variety, the amount of seed, and an accession number. Another system is to have two books, one for seeds coming in and one for seeds going out (Vernooy et al., 2020). Record as much or as little details as needed.

As a seed bank grows, it may be necessary to keep more detailed records for accessions, suppliers, seed multiplication, seed distribution, finances, and germination trials.

Computer based systems

Computer based systems also vary from simple to complex. Small seed banks can use a Microsoft® Excel sheet for inventory on hand and for transactions. Google documents available through Gmail provide easy accessibility through Wi-Fi, rather than only being stored on a local device or computer, and they can be shared with others or restricted. Inventory apps are available on some phones.

Advanced computer databases, such as those that utilize barcodes, are expensive but necessary for large seed banks that deal with large amounts of inventory and consistent seed movement. With these advanced systems, it is easier to filter and interpret data and history through reports.

In simple or advanced systems, you eventually might need to create stock item numbers for items that share the same accession but are inventoried separately, such as bulk seed and packets.

Seed banks often use paper systems and computer systems in conjunction with one another. If paper records were initially used for receiving new accessions, you can later input information from these into a database and then scan them if desired. Redundancy is beneficial if either the paper or computer system becomes compromised.

Record keeping is crucial and is an asset to a seed bank in managing and preserving local biodiversity and ensuring that the community has access to locally adapted seeds.

Replenishment of Seed Stocks

Part of inventory management is restocking when seed inventory is low or when germination rates decline. Keep viable seeds of local varieties in stock by generating or purchasing new seeds. Seed multiplication depends on local demand and seed bank ability. Crops that are popular will be in the most demand to be regrown, but the seed bank's ability to regrow the crop depends on land, season, and resource availability.

Seed banks produce seed at the seed bank site or on members' land. Provide training on seed production and make visits to farmers. Seed banks could involve members in seed production by asking for volunteers or requiring members to grow at least one crop variety per year or per season. If demand and ability are present, produce marketable varieties on a large scale for sale.

In certain situations, purchasing seed from outside suppliers is more economical and time-saving. If feasible, this can be a good option when space to grow out plants is limited or when you want to avoid having to control pests in the field. Most seed banks, however, will need to grow some of their crops to replenish seed stocks.

When seeds are being grown out in the field, keep records of which seeds are planted in which plots, both on a paper map of plots and with markers in the field. Record the seed spacing and important dates like planting, flowering, and harvesting. Include other growth data such as how many plants had to be removed for inferior quality or disease and what farming practices were used. Record the timing and uniformity of the harvest and what the seed harvest weight or quantity was. You may also want to record information about how the seed was processed and stored for future reference. If several people are collecting data, ensure that the data ends up in the same place where all seed bank workers can access it. See Part 1 for more information on multiplying seeds.

Chapter 14. Packaging and Labeling

Packaging Options

Containers for distributing seed can be made of plastic, paper, mylar, cloth, or other materials. Packets are often used for sharing small amounts of seed for people to trial. They contain enough seed for a small plot or row, not an entire field. Bulk quantities of seed can be transported in grain bags. See chapter 11 for information on storage containers.

Proper Labeling

Labeling is a key part of organizing seeds and communicating to others what type of seeds are in a container. Seed packets, containers, and bags in the seed bank must be clearly marked with the accession number, common name, variety name, germination percentage, and date tested. This is not in place of a paper or electronic copy of this data but is in addition to it.

Bulk seed label

Other things that could be included on the label for bulk seed would be the year the seed was acquired and the source of the seed.

Labeling seeds for distribution

Seed that is being distributed should be clearly labeled for farmers and community members. Farmers will want to know the variety, year seed was harvested, origin, days to maturity, plant height and habit, fruit size, fruit shape, fruit color, flavor, storage qualities, disease or pest susceptibilities, planting rate, when to plant, and processing or preparation instructions. On the label, include a summary of these details.

An example of a label for distribution:

> *Phaseolus coccineus* – Scarlet Runner Bean - 16049FL-191A
>
> This edible bean originates from the highlands of Central America. The vines grow rapidly and produces showy orange flowers. It is perennial if not damaged by frost. 65 days to harvest. Plant 5cm deep, 20cm apart along a trellis. Beans are edible and delicious when young or can be harvested when dry. Harvest frequently for continuous blooming.

Labels could also include local drawings of the plant variety, so that farmers know what the crop looks like.

Seed Distribution

Make seeds available in small, trial packets, or in larger amounts, depending on the goals of the seed bank. Small packets are easy to transport and can be used by seed recipients to evaluate multiple crops on small plots of ground. Distributing larger quantities may not be practical for seed banks that feature many crops/varieties. However, consider offering bulk quantities of a few varieties that are especially needed and valued within a community and that cannot be obtained elsewhere.

Distribution

Ensure consistency and fairness by filling packets or containers with consistent weights (measure with scale) or volumes (using teaspoons, tablespoons, cups etc.) of seed. Disseminate seeds in containers or plastic wrapping that excludes humidity. Vacuum sealed plastic bags (especially bags that can be custom sized using the vacuum sealer) work well. Where there is no electricity for a vacuum sealer, look for commonly used containers that can be re-used and sealed. All seed containers must be clearly marked with the accession number and name.

Supplemental handouts can also be printed to provide information such as days to maturity, plant height and habit, fruit traits, storage qualities, disease or pest susceptibilities, planting rate/seed spacing, when to plant, and processing or preparation instructions.

Part V. Seed Viability

Chapter 15. Testing Seed Viability

The Importance of Testing Seeds Regularly

To ensure that recipients are receiving high quality, viable seeds, periodically test seeds stored in your seed bank. Testing should be done on a yearly basis, systematically testing each lot of seeds throughout the year. Seed viability is a measure of the percentage of seeds that are alive after storage. The greater the viability of the seeds, the fewer the seeds that will be needed to establish a desired number of plants in the field or nursery.

You can test seed viability in many easy ways, but the most common method is a simple seed germination test. In a germination test, seeds receive the resources needed to germinate and grow into a seedling, including moisture, warmth, oxygen, and light.

A low amount of germination in a particular lot of seeds indicates that the seeds need to be regrown and replaced, while a high amount of germination confirms that the seed lot can remain in storage until the following year when they are tested again. Germination rates are also very useful to the farmers that will plant the seeds, as they might gain an accurate picture of how many seeds need to be planted in order to receive the desired outcome they seek. For this reason, germination percentage needs to be calculated and labeled clearly on seed packets before distribution.

Calculating Germination Percentage (%)

We recommend using a batch of 50 seeds per germination test to calculate germination percentage (%) because it is easy to calculate a percentage out of 100. Using more than 50 seeds can be wasteful and unnecessary. If working with a low number of seeds, then 25 seeds should suffice, but avoid using any less than 25 to retain accuracy.

When testing a batch of 50 seeds, double the final number of germinated seeds to determine the germination percentage. Example: If 45 out of 50 seeds germinate then the final germination percentage will be 90% (twice the number of viable seeds). If 20 out of 50 seeds germinate, the germination percentage (%) will be 40%.

Seed Viability Testing
Seed Germination Data Collection and Calculation

				Days After Beginning the Test								
Replication	Seed Type	Starting Date	Number of Seeds	2	4	6	8	10	12	14	Total Germinated	Germination Rate (%)
1	Lablab	20-Jun	50	16	10	9	5	3	2	0	45	90

After placing seeds into the germination chamber, check them every two days. Record and remove germinated seeds to prevent the growth of fungus. Add small amounts of water if seeds appear to be drying out. Tests should be moist, but never flooded.

Sanitation Practices

When setting up a germination test, keep everything clean at all times. Remember that the ideal conditions provided for seeds to germinate are the same conditions that fungus and bacteria thrive in. Keep hands, tools, workspaces, and seeds clean to mitigate the introduction of fungal spores that may negatively affect the germination test. If available, use alcohol (70% isopropanol or ethanol) to wipe down the workspace and any tools you might use, such as tweezers or containers. Consider using diluted hydrogen peroxide or diluted household bleach to sanitize seeds in preparation for germination testing. Household bleach is often between 5% and 6% sodium hypochlorite by volume. Dilute household bleach to 10% (between 0.5% and 0.6% sodium hypochlorite) with distilled water. Soak seeds for 5 minutes in diluted bleach solution. Thoroughly rinse seeds 3 times in distilled water after the treatment (Gilbert et al., 2023).

Germination testing supplies include (left) a bottle of rubbing alcohol (isopropanol) for sterilizing tools and equipment, (middle) a bottle of diluted bleach solution, and (right) a bottle of clean water (deionized).

Germination Test Methods

Germination tests can be carried out through different methods. To obtain the best results, match the seed you intend to germinate with a method that mimics (in the simplest way possible) the seed's natural germinating environment. Important factors to consider include space, moisture level, air flow, light exposure, etc. Refer to the Seed Crop Reference Table in the appendix for suggested germination methods based on crop species. You can also trial different methods with the same seed type to see what works best in your context. Note: some seed types need scarification (e.g. nicking, abrasion), presoaking, or temperature treatments to exit dormancy and germinate.

We will cover three common germination test methods: the ragdoll method, the petri dish method, and the direct seeding method.

The 'Ragdoll' Method

Step 1. Thoroughly rinse the seeds with clean water. A diluted solution of bleach (5%) can be used to kill off any fungal or bacterial colonies living on the seed.

Step 2. Pour the contents through a sieve. Rinse again with clean water if bleach was used in step one.

Step 3. Lay out one to two pieces of paper towel and wet the paper towel(s) using clean water. Wet the towel thoroughly but do not saturate it to the point of dripping.

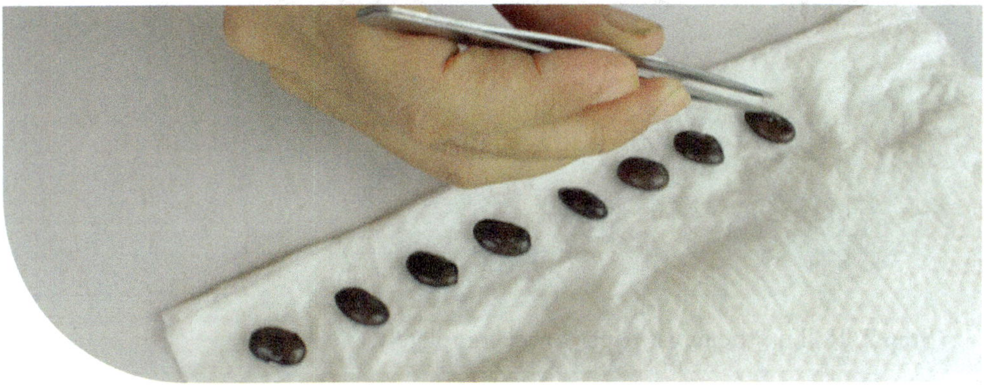

Step 4. Carefully place the seeds onto the paper towel in evenly spaced rows. Use tweezers to avoid contact with your hands, if possible, to help maintain cleanliness. Count the seeds to ensure that you have the desired number.

Step 5. Roll up the paper towel to cover all seeds. If you must use your hands at this stage, wash or disinfect them beforehand.

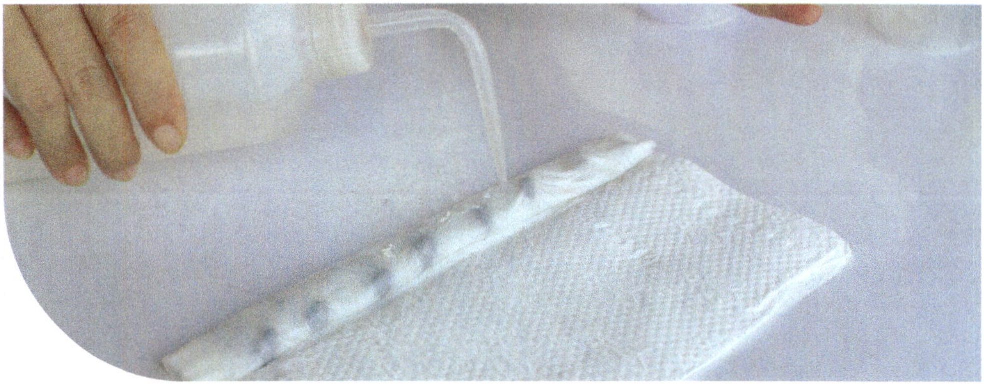

Step 6. Wet the paper towel as you roll it up to ensure that seeds have adequate moisture. Saturate the paper towel but not to the point of dripping.

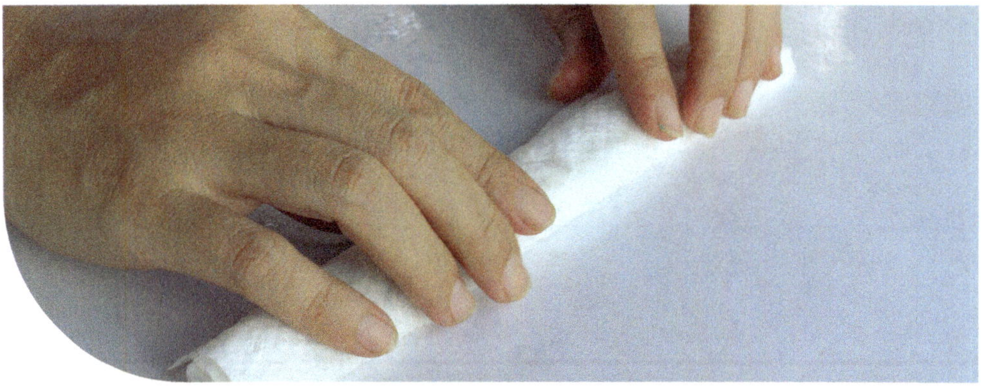

Step 7. Roll up the moist paper towel completely.

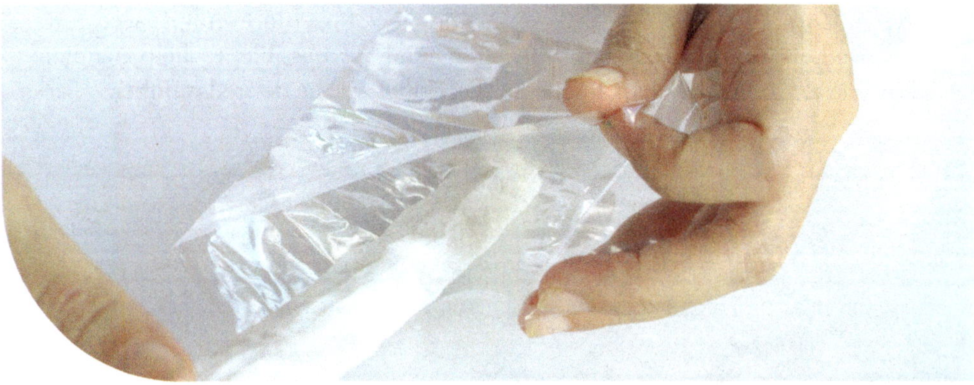

Step 8. Place the entire 'ragdoll' into a plastic bag to retain moisture. The bag also protects the seeds from additional fungal spores in the air.

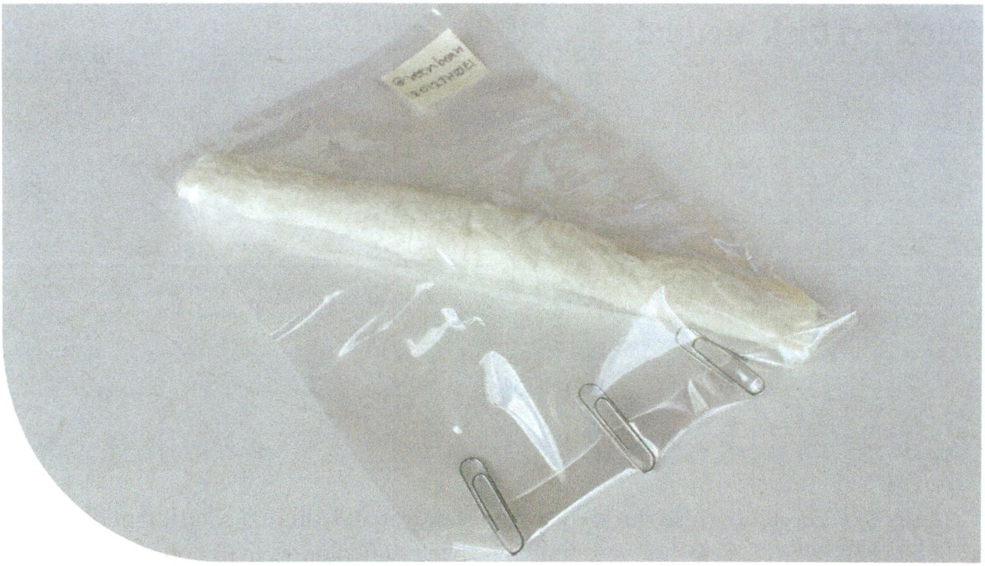

Step 9. Partially seal the bag so that it still 'breathes.' Label the bag with the date, crop name, and accession.

Step 10. Open the rag every two days to check how many seeds have germinated. Record the number of germinated seeds, then remove the germinated seeds and reseal the towel. If any seeds are moldy or rotting, remove them as well. Place it back into the germination chamber.

The Petri Dish Method

Step 1. Thoroughly rinse the seeds with clean water. A diluted solution of bleach (5%) can be used to kill off any fungal or bacterial colonies living on the seed.

Step 2. Pour the contents through a sieve.

Step 3. Use clean water to thoroughly rinse the seeds again if bleached was used in step one.

Step 4. Prepare the petri dish by placing a sheet of paper towel on the bottom and wetting it with clean water. Wet the towel thoroughly but do not saturate it to the point of pooling.

Step 5. Carefully place the seeds onto the towel in evenly spaced rows. Use tweezers to avoid contact with your hands, if possible, to help maintain cleanliness. Count the seeds to ensure that you have the desired number.

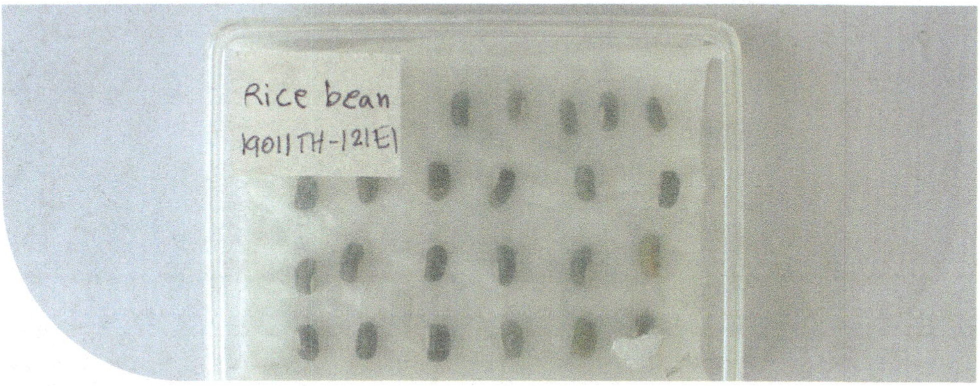

Step 6. Close the top of the petri dish with the lid. An enclosed container ensures that moisture stays in and does not escape and protects the seeds from fungal spores in the air. Label the dish with the date, crop name, and accession.

Step 7. Open the petri dish every two days to check how many seeds have germinated. Record the number of germinated seeds, then remove the germinated seeds and reseal the container. If any seeds are moldy or rotting, remove them as well. Place the dish back into the germination chamber.

Step 8. If seeds are too moist, or have not been thoroughly cleaned, they may be susceptible to fungal disease which can negatively impact the test. To avoid this scenario, pay careful attention to cleanliness and sanitation.

The Direct Seeding Method

Step 1. Prepare a seed tray with enough plugs for each seed to be planted individually. Use a well-draining potting mix to ensure that seeds do not sit in standing water.

Step 2. Place one seed in each plug and cover with soil.

Step 3. Once the seeds have been planted, label the seed tray with the crop name, accession, and date. If space allows, seed more than one batch of seeds per tray, with clear separation and labels.

Step 4. Check trays for germination every two days. Once seeds germinate and break the soil surface, record them as germinated. Keep all seedlings in place until the germination test is complete so there is no confusion as to whether a seed has germinated. Note: this tray has two separate batches of germination tests happening at once.

Part VI. Seed Sharing & Distribution

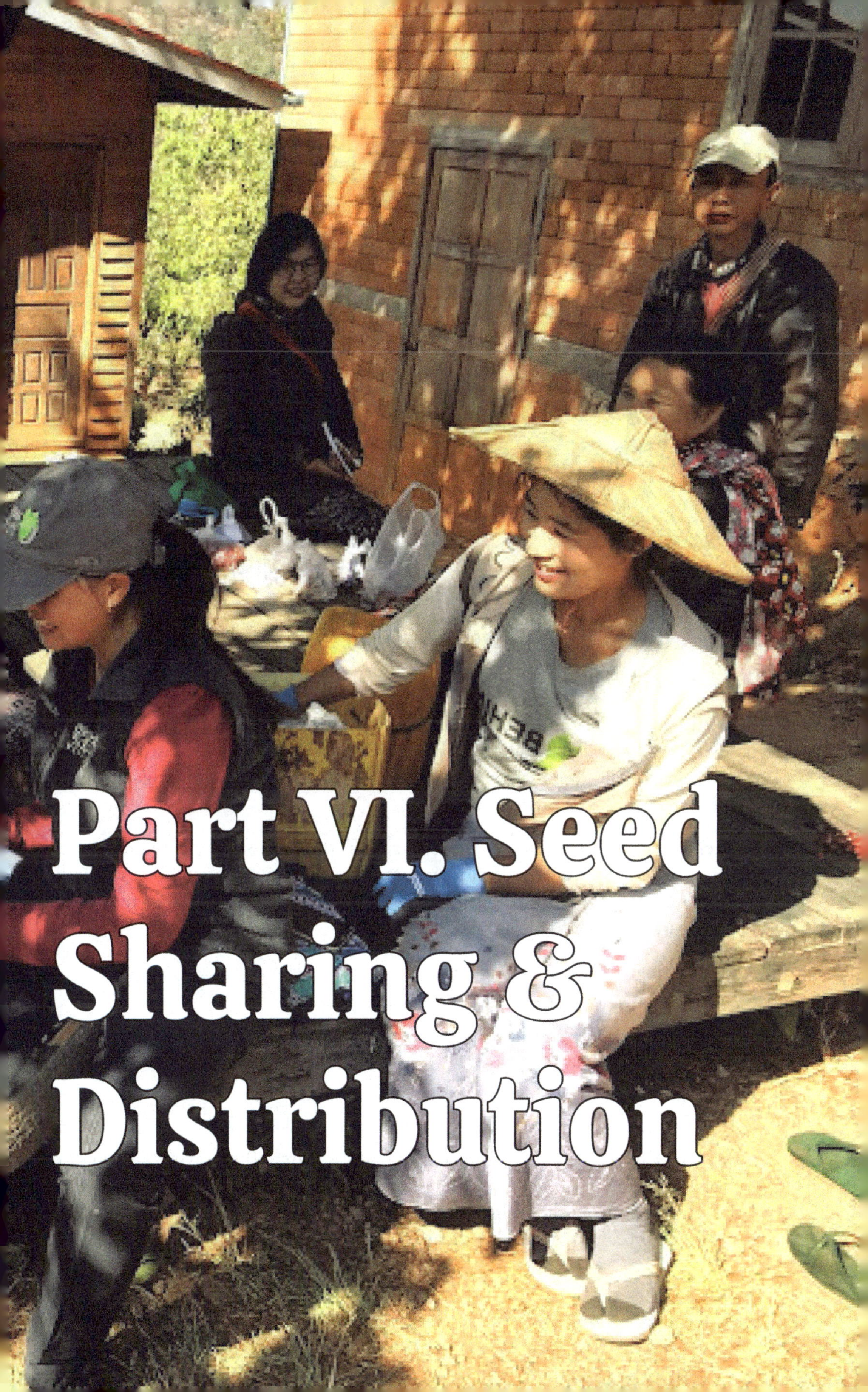

Chapter 16. Seed Exchanges

An ECHO conference seed exchange

An Introduction to Community Seed Exchanges

Adapted from: *ECHO Technical Note* 67 (Bakker and Zenén, 2011)

Historically, community markets have helped provide farmers with replacements for seeds that were lost due to unproductive crops, plagues, or natural disasters. Farmers also search markets for seeds that are better appreciated, for taste or production traits, than the ones they are already cultivating.

Seed exchanges (also referred to as "seed swaps" or "seed fairs"), with methodical planning and careful selection of participants, provide greater potential of successful seed distribution than traditional markets. Organizers of seed exchanges may try to meet one or more of the following objectives:

- Improve farmer access to seeds
- Maintain or improve agrobiodiversity
- Raise awareness of the importance of biodiversity and plant genetic resources GR)
- Strengthen the rights and sovereignty of farming communities
- An Important Source of Seeds for Farmers (and CSBs)

Farmers typically acquire seeds (and other planting material) by saving a portion of their harvest; borrowing, exchanging, or bartering with other farmers; buying seeds in the market; or acquiring them through extension agencies. In a development context, most seeds are produced by farmers themselves rather than by plant breeders.

An ECHO organized community seed exchange

Smallholder farmers do not typically restrict themselves to seeds from one source, and often use both local varieties (or landraces) and seeds originating from the institutional plant breeding system. What seed source they use depends on the crop, use of the seed (e.g. for sale or for home consumption), availability (often markets do not have the required seeds at the right time) and accessibility (seeds produced by plant breeders can be too expensive for some farmers).

Seed exchanges that are concerned with conserving agro-biodiversity and promoting landraces came about as a response to the loss of diversity and the realization that modern varieties do not always adapt to the farming systems of small-scale farmers. A second type of seed exchange emerged as an attempt to provide alternative seed options (other than handouts) to farmers affected by devastating crop losses. This second type of exchange usually makes use of a voucher system to allow even the poorest farmers access to seed.

By growing their diverse crops, farmers are actively conserving both diversity and the specific knowledge required for appreciating and maintaining that diversity. Diversity needs to be applied to be useful. If the wealth of seed diversity were confined to seed banks, it would be of little use. By creating a special occasion (i.e. a seed exchange), farmers can access each other's seed. At the same time, they contribute to a greater awareness of the importance of diverse PGR. Finally, seed exchanges bring different stakeholders in PGR together.

How to Organize and Facilitate a Seed Exchange

Adapted from: *ECHO Asia Note* 16 (Tshin, 2013)

ECHO Asia Impact Center staff have developed and facilitated seed exchange events during meetings with local partners for many years. Through conversations with farmers and NGO staff, ECHO has gained better insight into important plant varieties, seed saving practices, attitudes towards saving seeds, and sustainable farming. The most exciting part of these events is hearing success stories! At the first ECHO Asia agriculture workshop in Myanmar in 2012, one organization from southern Shan State brought over 75 cuttings of Chaya to share with delegates, after having successfully introduced it as a food and forage plant on their small farm resource center. Their stock came from a single cutting received only one year earlier during ECHO Asia's biennial Tropical Agriculture conference in Chiang Mai!

Here we share some methods for implementing seed exchanges that can be carried out in local languages and attended by local farmers and development workers. The materials required for a seed exchange are basic, but the process of exchanging and capturing information requires prior preparation and adjustments based on the culture and realities in the community.

We recommend reading the article "Seed fairs: Fostering local seed exchange to support regional biodiversity" (Meitzner and Ricciardi, 2012) as it explains some of the reasons it is important to build capacity around local plant biodiversity and understand local seed systems.

Basic materials for exchanging seeds:

- Enough seeds and plant propagation material to share
- Small re-sealable bags in which to distribute seeds

- In Thailand, we buy small plastic medicine bags from local pharmacies
- You can easily substitute newspaper or sheets of paper, paper envelopes, or bags

- Permanent markers and labels
  - To clearly identify seed packages with common and variety names (preferably in local language as well as in English)

Basic materials for the seed exchange facilitator:

- White board or poster paper to list regions or organizations in attendance
- Clipboards, pens, and copies of a simple data sheet for collecting seed information
- Local language translators
- People ready to record seed information

Steps to Consider when Planning a Seed Exchange event:

1. Well before the event, communicate with and encourage participants to bring seeds and plant propagation material to the meeting.

- Include requests for seeds and plant material when you begin advertising the meeting with local organizations. If possible, request each organization to bring a minimum of 500 grams of seed to be able to share with attendees, as well as materials that help explain special characteristics of each species (for example, large color photographs, seedlings, videos, etc.). Consider that adequate lead time is needed for delegates to gather seeds and plant materials to share.

2. Decide on parameters of seed exchange.

- If you are conducting a multi-day meeting/workshop, consider: 1) which days would be best for collecting seed information, 2) if discussions among delegates are needed, and 3) how to conduct the exchange activity (see below). You may need to set a deadline during the workshop for accepting seeds into the exchange.
- Consider who should receive seeds. Should delegates receive shared seeds if they did not bring any to share? Or should only delegates who bring seeds participate in the exchange? ECHO Asia typically brings additional seeds for all delegates to receive as a "gift bag" whether they brought any to share or not.
- Consider the amount of seeds brought to the meeting – there may or may not be enough to share with everyone. Should every delegate from every organization receive seeds? Or should each organization receive only one set of seeds?

3. Determine method for implementing seed exchange during the meeting.

- Meetings that take place over several days are ideal because it takes time to collect seed information from delegates before the actual exchange. You may need to assign one person to manage and prepare for the exchange apart from regular meeting events. Consider what time constraints you may be faced with, because preparations and the exchange activity may take more time than anticipated.

Setting Up a Seed Exchange:

1. Register the seeds brought by delegates separately from the normal registration process of the main training event.

ECHO Asia "registers" the seeds to compile a master list of varieties shared at our meetings and to record indigenous plant and seed saving information. Typically, this process requires translators to ask questions in the local language and people to record information in Thai or English. A Seed Exchange Template Form has been provided below.

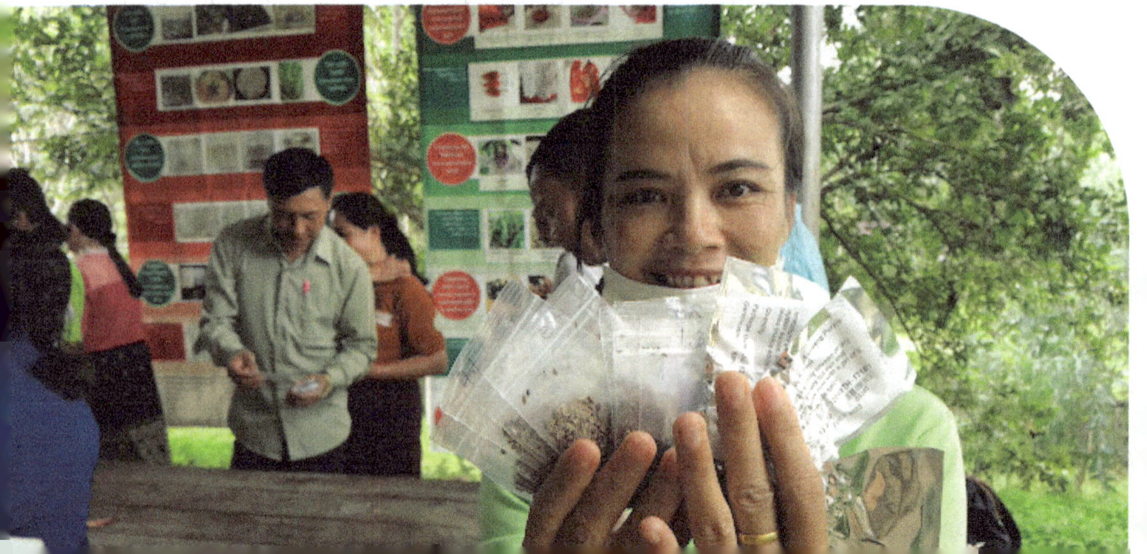

ECHO Asia's Seed Source information

(Before we place seeds in the seed bank and access them we need the following information. Please take some time to complete the form and provide as much detailed information as possible.)

<table>
<tr><th>Seed Source Information</th><th>Plant Characteristics</th></tr>
<tr><td>ECHO staff in involved in transfer and Date of Transfer
วันที่ย้ายส่งเมล็ด</td><td rowspan="4">Family Name:
ตระกูล

Scientific Name of plant:
ชื่อวิทยาศาสตร์
Common name
ชื่อสามัญ

Variety
พันธุ์พืช

Description of plant: ลักษณะของพืช
Seed color สีของเมล็ด

Features of main parts harvested (pod/fruit/shoots?)
ลักษณะเมล็ดที่เก็บเกี่ยว(กอม,ฝัก,ลูก)

Edible? เป็นอาหารทานได้

Marketing/Can you sell it? การตลาด/ขายอย่างไร?

Where are seeds usually planted? In garden? Forest?
ปลูกในสวนครัว?ปลูกในป่า?

What sets this variety apart of other?ลักษณะพิเศษของพืช?

How does it look/test/cook/store differently from other?
สังเกตโดยวิธีไหน/รสชาติ/ทำอาหาร/การจัดเก็บแตกต่างจากอย่างอื่นอย่างไร?</td></tr>
<tr><td>Source Name
ชื่อของบุคคลที่ให้เมล็ด:

Source Address:
ที่อยู่ของบุคคลที่ให้เมล็ด

Source Telephone/Email:
เบอร์โทรศัพท์บุคคลที่ให้เมล็ด

The seeds were (circle one):
Free / Purchased ____________(price)
ฟรี / ซื้อ เท่าไร ____________ บาท</td></tr>
<tr><td>Place where seeds originated
(village หมู่บ้าน/ district อำเภอ country จังหวัด):

Additional Information on Seed Source:
ข้อมูลเพิ่มเติม</td></tr>
</table>

2. Prepare seeds and plant material for the exchange.

Delegates may bring seeds in bulk amounts that then need to be split into smaller portions and into re-sealable bags. For seeds like beans, measure out 10-20 seeds per small bag. For smaller seeds like amaranth or mustard green, measure out 1 tablespoon into the smaller bags. Be prepared for late seed submissions after the registration process is complete. Clearly label bags containing seeds with relevant information and clearly mark and display seed names neatly on tables or on the floor where they will be placed.

3. Implement the actual seed exchange: 45 to 60 minutes total duration.

Give each organization 5-10 minutes to introduce their seeds and why they are sharing them with the group. Possible sharing points include:

- showing actual seedlings,
- showing large color photos of lesser-known plant varieties,
- explaining growing and seed harvesting methods, and
- explaining the use of each plant.

Prepared seed factsheets are also a great way to disseminate information. A timekeeper is important to keep this part of the exchange moving along. Following introductions, the exchange activity usually takes 15-30 minutes to complete. However, the exchange format depends on how many delegates are in attendance:

For 100 or fewer people, call groups one at a time to collect a limited set (one set of seeds per group) or a designated amount of seeds. This helps set an efficient pace. Give priority to groups that brought seeds to exchange and allow them to choose seeds first. If there is a sufficient amount of seeds, consider multiple selection rounds for each group.

For our biennial conference of over 100 delegates, a more methodical process is required. We encourage delegates to pre-order seeds and assign a team to manage distribution throughout the duration of the conference.

ECHO
SEED VIABILITY TESTING
ECHO

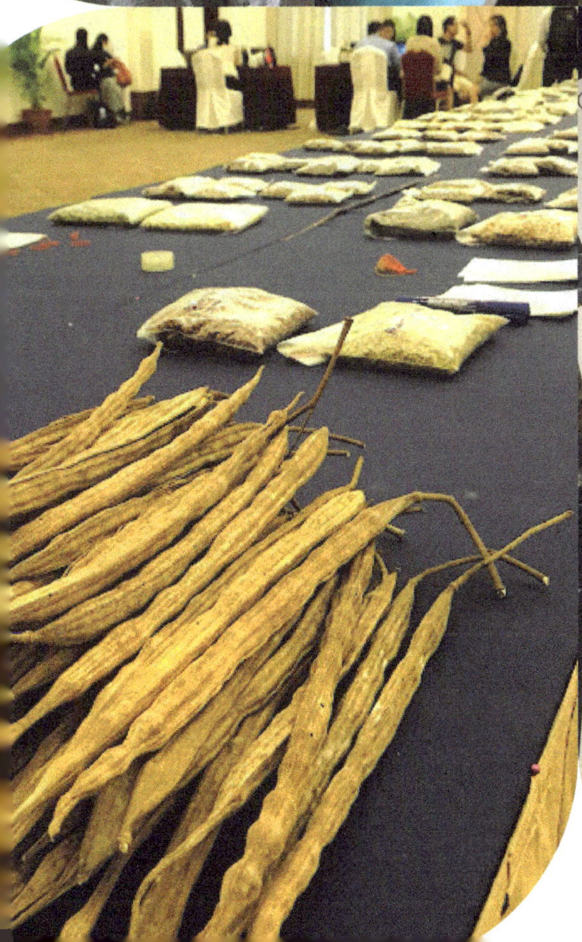

Moringa
Eat leaves like spinach
asparagus. Grate root
seeds for oil. Plant do
Extremely nutritious w
potassium, calcium,
acids. Also highly m
Packaged: 201
Germination:
Clitoria ternatea
Butterfly Pea
Perennial legume. Likes
drought-tolerant legume
Small leaves; not a thic
used as dye.
Packaged: May
Germination:
Phaseolus lunatus
Lima Bean
1402
Seeds are known for their sweetness and
cook. Plant 4cm deep in well-drained soil
a green manure and cover crop. Fixes nit
trellised or not.
Packaged: 2014
Germination:

Appendix: Seed Crop Reference Table

Common Name(s)	Scientific Name	Planting Season	Pollination Type	Minimum Isolation Distance(m)	# of plants needed for pollination to occur	# of plants needed for Variety Maintance	Seed Cleaning Methods	Seed Germination method(s)	Acceptable Perentage Germinaton
Adzuki Bean , Red Bean	*Vigna angularis*	Warm	Self-pollinat-ed	3-6	1	10-25	dry pod	rag	75
African Marigold	*Tagetes erecta*	Warm	Cross-polli-nated, insect				dry pod	direct seed	80
Amaran-th(grain)	*Ama-ranthus Cruentus*	Warm, wet	Cross-polli-nated, insect	153	1	5-25	mortor	petri	60
Amaran-th(vegeta-ble)	*Ama-ranthus tricolor*	Warm, wet	Cross-polli-nated, insect	153	1	5-25	mortor	petri	60
Angeled Luffa	*Luffa acutangula*	Warm, wet	Cross-polli-nated, insect	805 (1/2 mile)	1	5-10	dry pod	direct seed	60
Bitter Gourd	*Momordica charantia*	Any	Self-pollinat-ed	400	1		sand and screen	direct seed	60
Buckwheat	*Fagopyrum esculentum*	Cool	Self or Cross-polli-nated, variety dependent	244-805	5	20-50	dry pod	direct seed	75
Butterfly Pea	*Clitoria ternatea*	Warm	Self-pollinat-ed, can cross via insects		1		dry pod	petri / rag	60

Common Name(s)	Scientific Name	Planting Season	Pollination Type	Minimum Isolation Distance(m)	# of plants needed for pollination to occur	# of plants needed for Variety Maintance	Seed Cleaning Methods	Seed Germination method(s)	Acceptable Perentage Germinaton
Calabash Gourd	*Lagenaria siceraria*	Warm, wet	Cross-pollinated, insect	3-6	1	5-10	fermentation	direct seed	60
Calendula	*Calendula officinalis*	Cool	Cross-pollinated, insect				dry pod	direct seed	
Carrot	*Daucus carota*	Cool	Cross-pollinated, insect	244-805	5	20-50	gentle hit	direct seed	55
Cassod Tree	*Senna siamea*	Warm	Cross-pollinated, insect		1		dry pod	direct seed, scarify/ soak first	10
Chaya	*Cnidoscolus aconitifolius*	Any	Vegetative propogation, cuttings	N/A	N/A	N/A	N/A	N/A	N/A
Chia	*Salvia hispanica*	Cool	Self-pollinated, can cross via insect				mortar or gentle hit	petri	60
Chick Pea	*Cicer arietimum*	Cool, wet	Self-pollinated, can cross via insect	805 (1/2 mile)	1	5-10	dry pod	rag	60
Chili Pepper	*Capsicum frutescens*	Warm, wet	Self-pollinated, can cross via insect	153	1	5-20	hit or massage	direct seed	55
Chinese Kale	*Brassica oleracea var. alboglabra*	Cool, wet	Cross-pollinated, insect	Time isolation	5	20-50	gentle hit	petri	60

Common Name(s)	Scientific Name	Planting Season	Pollination Type	Minimum Isolation Distance(m)	# of plants needed for pollination to occur	# of plants needed for Variety Maintance	Seed Cleaning Methods	Seed Germination method(s)	Acceptable Perentage Germinaton
Choy Sum	*Brassica rapa var. parachin-ensis*	Cool, wet	Cross-polli-nated, insect	1608 (1 mile)	5	20-50	gentle hit	petri	75
Coriander/ Cilantro	*Cori-andrum sativum*	Cool	Cross-polli-nated, insect	805 (1/2 mile)	5	20-50	hit or massage	petri / rag	60
Cowpea	*Vigna unguicula*	Warm, wet	Self-pollinat-ed, can cross via insect	3-6	1	10-25	dry pod	rag	75
Cranberry Bean	*Phaseolus vulgaris*	Warm	Self-pollinat-ed, can cross via insect	10	1		dry pod	rag	75
Cranberry Hibiscus	*Hibiscus acetosella*	Warm	Self-pollinat-ed, can cross via insect	1	1		dry pod	rag / direct seed	60
Cucumber	*Cucumis sativus*	Warm	Cross-polli-nated, insect	805 (1/2 mile)	1	5-10	fermanta-tion	direct seed	80
Dill	*Anethum graveolens*	Cool, dry	Cross-polli-nated, insect	Time isolation or 244-805	5	20-50	gentle hit	direct seed	60
Eggplant	*Solanum melongena*	Warm, wet	Self-pollinat-ed, can cross via insect	91-488	1	5-20	massage or hit	petri	60

Common Name(s)	Scientific Name	Planting Season	Pollination Type	Minimum Isolation Distance(m)	# of plants needed for pollination to occur	# of plants needed for Variety Maintance	Seed Cleaning Methods	Seed Germination method(s)	Acceptable Perentage Germinaton
Ethiopian Kale	*Brassica carinata*	Cool	Cross-pollinated, insect				gentle hit	petri	60
Fennel	*Feoniculum vulgare*	Cool, wet	Cross-pollinated, insect	244-805	5	20-50	gentle hit	direct seed	60
Field Lablab	*Lablab purpureus*	Warm, wet	Self-pollinated, can cross via insect		1		dry pod	rag/direct seed	60
Flemingia	*Flemingia macrophylla*	Warm, wet	Cross-pollinated, insect				dry pod	direct seed, scarify/soak first	10
Flint Corn	*Zea mays var. indurata*	Warm, wet	Cross-pollinated, wind	244-805	10	50-120	dry cob	rag	75
Foxtail Millet	*Setaria italica*	Warm, wet	Self-pollinated				mortar or gentle hit	petri/direct seed	60
French marigold	*Tagetes patula*	Warm	Cross-pollinated, insect				dry pod	direct seed	80
Ghost Pepper	*Capsicum chinense*	Warm	Self-pollinated, can cross via insect	153	1	10	hit or massage	direct seed	55
Green Bean	*Phaseolus vulgaris*	Warm, wet	Self-pollinated	10	1	5-10	dry pod	rag	75
Holy Basil	*Ocimum tenuiflorum*	Warm	Cross-pollinated, insect				mortar	direct seed	60

Common Name(s)	Scientific Name	Planting Season	Pollination Type	Minimum Isolation Distance(m)	# of plants needed for pollination to occur	# of plants needed for Variety Maintance	Seed Cleaning Methods	Seed Germination method(s)	Acceptable Perentage Germinaton
Horse Gram	*Macrotyloma uniflorum*	Warm, dry	Self-pollinated		1		dry pod	direct seed, scarify/ soak first	60
Hummingbird Tree	*Sesbania grandiflora*	Warm, wet	Cross-pollinated, bird				dry pod	direct seed	10
Inca nut	*Plukenetia volubilis*	Warm, wet	Cross-pollinated, insect				dry pod	direct seed	60
Indian Trumpet Flower	*Oroxylum indicum*	Warm, wet	Cross-pollinated, bat				dry pod	direct seed	
Jack Bean	*Canavalia ensiformis*	Warm, dry or wet	Self-pollinated, can cross via insect	805 (1/2 mile)	1	5-10	dry pod	ragdoll	60
Job's Tears	*Coix lacryma-jobi*	Warm, wet	Cross-pollinated, wind				dry pod	direct seed	75
Large Indigo	*Indigofera zollingeriana syn.L. teysmannii*	Warm, wet	Cross-pollinated, insect				mortar	direct seed	60
Lettuce	*Lactuca sativa*	Cool, wet	Self-pollinated, can cross via insect	3-6	1	5-10	gentle hit	petri	60
Lima Bean	*Phaseolus lunatus*	Cool, wet	Self-pollinated, can cross via insect	49-152	1	10-25	dry pod	rag	70

Common Name(s)	Scientific Name	Planting Season	Pollination Type	Minimum Isolation Distance(m)	# of plants needed for pollination to occur	# of plants needed for Variety Maintance	Seed Cleaning Methods	Seed Germination method(s)	Acceptable Perentage Germinaton
Malabar Spinach	*Basella alba*	Warm, wet	Self-pollinat-ed	N/A (out-crossing rare)	1		sand and screen	direct seed	60
Moon-flower	*Ipomoea alba*	Warm	Cross-polli-nated, insect				dry pod	rag/direct seed	60
Moringa	*Moringa oleifera*	Warm, dry	Cross-polli-nated, insect				dry pod	direct seed	60
Mung Bean	*Vigna radiata*	Warm, dry	Self-pollinat-ed	N/A (out-crossing rare)	1	20-50	dry pod	rag/petri	60
Musk Melon	*Cucumis melo*	Warm, wet	Cross-polli-nated, insect	244-805 (1/2 mile)	1	10	fermenta-tion	direct seed	80
Mustard Greens	*Brassica juncea*	Cool, wet	Cross-polli-nated, insect	805 (1/2 mile)	5	20-50	gentle hit	petri	75
Naranjilla	*Solanum quitoense*	Cool, wet	Self-pollinat-ed, can cross via insect		1		fermenta-tion	petri	60
Nemati-cidal Mari-gold	*Tagetes minuta*	Warm	Cross-polli-nated, insect				dry pod	direct seed	80
Niger Seed	*Guizotia abyssinica*	Cool	Cross-polli-nated, insect				gentle hit	direct seed	70
Okra	*Abel-moschus esculentus*	Warm, dry or wet	Cross-polli-nated, insect	1608 (1 mile)	1	5-10	dry pod	ragdoll	50

Common Name(s)	Scientific Name	Planting Season	Pollination Type	Minimum Isolation Distance(m)	# of plants needed for pollination to occur	# of plants needed for Variety Maintance	Seed Cleaning Methods	Seed Germination method(s)	Acceptable Perentage Germinaton
Pak Choi/ Bok Choy	*Brassica rapa var. chinensis*	Cool, wet	Cross-polli-nated, insect	244-805	5	20-50	gentle hit	petri	60
Papaya	*Carica papaya*	Warm, wet	Self or Cross-polli-nated, variety dependent	500	1		sand and screen	direct seed	60
Passion Fruit	*Passiflora edulis*	Warm, wet	Self or Cross-polli-nated, variety dependent				sand and screen	direct seed	70
Pearl Millet	*Pennisetum glaucum*	Warm	Cross-polli-nated, wind	200-400			gentle hit	direct seed	70
Pigeon Pea	*Cajanus cajan*	Warm, wet	Self-pollinat-ed, can cross via insect	805 (1/2 mile)	1		dry pod	ragdoll	60
Popcorn	*Zea mays var.everta*	Warm, wet	Cross-polli-nated, wind	244-805	10	50-120	dry cob	direct seed	75
Quinoa	*Chenopodi-um quinoa*	Cool	Self-pollinat-ed	30-91	1	10-25	mortar	direct seed	60
Red Leu-caena	*Leucaena diversifolia*	Any	Self-pollinat-ed, can cross via insect				dry pod	direct seed, scarify/ soak first	10
Rice	*Oryza sativa*	Warm, wet	Self-pollinat-ed	3-6	1	5-10	hit	petri	80

Common Name(s)	Scientific Name	Planting Season	Pollination Type	Minimum Isolation Distance(m)	# of plants needed for pollination to occur	# of plants needed for Variety Maintance	Seed Cleaning Methods	Seed Germination method(s)	Acceptable Perentage Germinaton
Rice Bean	*Vigna umbellata*	Warm, wet	Self-pollinat-ed		1		dry pod	rag/petri	60
Roselle	*Hibiscus sabdariffa*	Warm, wet	Self-pollinat-ed, can cross via insect	805 (1/2 mile)	1		dry pod	rag/direct seed	60
Sesame	*Sesamum indicum*	Warm	Self-pollinat-ed, can cross via insect	50-100	1		dry pod	petri	60
Smooth Luffa	*Luffa ae-gyptiaca*	Warm, wet	Cross-polli-nated, insect	805 (1/2 mile)	1	5-10	dry pod	direct seed	60
Snake Gourd	*Tricho-santhes cucumerina*	Warm, wet	Cross-polli-nated, insect	500-1000	3		sand and screen	direct seed, scarify/ soak first	60
Snow Pea	*Pisum sa-tivum var. macrocar-pum*	Cool, wet	Self-pollinat-ed	3-6 (out-crossing rare)	1	5-10	dry pod	rag	80
Sorghum	*Sorghum bicolor*	Warm, wet	Self-pollinat-ed, can cross via wind and insect	30-61	1	10-25	gentle hit	direct seed	60
Soybean	*Glycine max*	Warm, wet	Self-pollinat-ed	N/A (out-crossing rare)	1	5-10	dry pod	rag/direct seed	75

Common Name(s)	Scientific Name	Planting Season	Pollination Type	Minimum Isolation Distance(m)	# of plants needed for pollination to occur	# of plants needed for Variety Maintance	Seed Cleaning Methods	Seed Germination method(s)	Acceptable Perentage Germinaton
Stylo	*Stylosan-thes capita*	Any	Self-pollinat-ed, can cross via insect				mortar or gentle hit	direct seed	60
Sunflower	*Helianthus annuus*	Warm, wet	Cross-polli-nated, insect	805-4824 depending on population	5	20-50	dry pod	rag/direct seed	60
Sunn Hemp	*Crotalaria juncea*	Warm, wet	Self-pollinat-ed, can cross via insect				dry pod	petri	60
Sweet Basil	*Ocimum basilicum*	Warm	Cross-polli-nated, insect	46			mortor	direct seed	60
Sweet Corn	*Zea mays var.saccha-rata*	Warm, wet	Cross-polli-nated, wind	244-805	10	200	dry cob	rag/direct seed	75
Sword Bean	*Canavalia gladiata*	Warm, wet	Self-pollinat-ed		1		dry pod	direct seed	60
Tepary Bean	*Phaseolus acutifolius*	Warm, dry	Self-pollinat-ed	3-6	1	5-10	dry pod	rag/petri	60
Tomato	*Solanum lycopersi-cum*	Warm	Self-pollinat-ed	3-15	1	5-10	fermenta-tion	direct seed	75
Vegetable Lablab	*Lablab purpureus*	Warm, wet	Self-pollinat-ed, can cross via insect		1		dry pod	rag/direct seed	60
Velvet Bean	*Mucuna pruriens*	Warm, wet	Self-pollinat-ed		1		dry pod	direct seed	60

Common Name(s)	Scientific Name	Planting Season	Pollination Type	Minimum Isolation Distance(m)	# of plants needed for pollination to occur	# of plants needed for Variety Maintance	Seed Cleaning Methods	Seed Germination method(s)	Acceptable Perentage Germinaton
Wax Gourd/ Winter-melon	*Benincasa hispida*	Warm, wet	Cross-polli-nated, insect	805 (1/2 mile)			sand and soil	direct seed	60
White Horay Pea	*Tephrosia candida*	Warm, wet	Self-pollinat-ed				dry pod	direct seed	60
White Leucaena	*Leucaena leucoceph-ala*	Any	Self-pollinat-ed, can cross via insect				dry pod	direct seed, scarify/ soak first	10
Winged Bean	*Psopho-carpus tetragonol-obus*	Warm, wet	Self-pollinat-ed				dry pod	rag	60
Yam bean/ Jicama	*Pachyrhi-zus erosus*	Warm, wet	Self-pollinat-ed		1		dry pod	direct seed, scarify first	60
Yard Long Bean	*Vigna unguicu-lata subsp. sesquiped-alis*	Warm	Self-pollinat-ed	N/A (out-crossing rare)	1	10-25	dry pod	rag	80
Zinnia	*Zinnia elegans*	Warm, dry	Cross-polli-nated, insect				mortar	direct seed	85

References

Ahn, J-E., X. Zhou, S.E. Dowd, R.S. Chapkin, and K. Zhu-Salzman. 2013. Insight into hypoxia tolerance in cowpea bruchid: metabolic repression and heat shock protein regulation via hypoxia-inducible factor 1. *PLoS One* 8(4), e57267 doi: 10.1371/journal.pone.0057267.

Almekinders, C. J. M. and N. P. Louwaars. 2002. The Importance of the Farmers' Seeds Systems in a Functional National Seed Sector. *Journal of New Seeds*. 4(1): 15-33.

Anwar, F., M. Ashraf, and M.I. Bhanger. 2005. Interprovenance variation in the composition of *Moringa oleifera* seeds from Pakistan. *Journal of the American Oil Chemists' Society*. 82(1):45-51.

Bakker, N. and F. Zenén. 2011. Farmers' Seed Fairs. *ECHO Technical Note* no. 67.

Bellon, M., J. Becerril, M. Adato, J. Alfonso, A. Gómez, and D. Mexico. 2003. The Impact of Improved Maize Germplasm on Poverty Alleviation: The Case of Tuxpeño-Derived Material in Mexico.

Bicksler, A. J. 2017. Building a Low-Cost Seed Dryer for Local Seed Banks. *ECHO Asia Notes* no. 31.

Bicksler, A. J. 2011. Testing Seed Viability Using Simple Germination Tests. *ECHO Asia Notes* no. 11.

Bunch, R. 1982. *Two Ears of Corn: A Guide to People-centered Agricultural Improvement*. World Neighbors.

Chala, M. and G. Bekana. 2017. Review on seed process and storage condition in relation to seed moisture and ecological factor. *Journal of Natural Sciences Research*. 7:84-90.

Conservation By Design Limited. 2012. MSDS Self-indicating green-orange silica gel. http://ge-iic.com/files/fichas%20productos/gel_silice_color.pdf.

Copeland, L.O. and M.B. McDonald. 2001. *Principles of Seed Science and Technology 4th Edition*. Springer Science-Buisness Media.

Croft, M., A. J. Bicksler, J. Manson, and R. Burnette. 2013. Comparison of appropriate tropical seed storage techniques for germplasm conservation in mountainous sub-tropical climates with resource constraints. *Experimental Agriculture*. 49(2): 279-294.

Croft, M., A. Bicksler, J. Manson, and R. Burnette. 2012. Vacuum sealing vs. refrigeration: Which is the most effective way to store seeds? *ECHO Asia Notes* no. 14.

ECHO Staff. 2016. Seed storage in the tropics. *ECHO Best Practice Note* no. 5.

El-Khier, K.S. Ishag, and A.E.A. Yagoub. 2008. Chemical composition and oil characteristics of sesame seed varieties grown in Sudan. *Research Journal of Agriculture and Biological Sciences*. 4(6):761-766.

FAO. 2014. Community seed banks. *Junior Farmer Field and Life School - Facilitator's guide.* Available at http://www.fao.org/3/a-i3987e.pdf

FAO. 1999. Women: users, preservers and managers of agrobiodiversity. *Women: The key to food security.* Available at https://www.fao.org/4/x0171e/x0171e03.htm#P181_22270

[FFIL] Feed the Future Innovation Lab (undated). Protocol for post-harvest seed drying and storage using zeolite desiccant beads. Available at https://horticulture.ucdavis.edu/information/protocol-postharvest-seed-drying-and-storage-using-zeolite-dessicant-beads (accessed 9 August 2022).

Frankland, B. and R.B. Taylorson. 1983. Light Control of Seed Germination. *Environmental Science, Biology.*

Ghimire, K.H. and B.K. Joshi. 2019. Crop groups based on conservation strategy. In: Working groups of agricultural plant genetic resources (APGRs) in Nepal (BK Joshi and R Shrestha, eds). Proceedings of a National Workshop, 21-22 June 2018, Kathmandu; pp.174-183.

Gilbert, G. S., A. Diaz, and H. A. Bregoff. 2023. Seed Disinfestation Practices to Control Seed-Borne Fungi and Bacteria in Home Production of Sprouts. *Foods.* 12(4): 747.

Global Seed Savers. n.d.. Seed Libraries. Retrieved August 23, 2022, from https://globalseedsavers.org/our-program/seed-library/

Guantian, L. and F. Zhu. 2018. Quinoa starch: structure, properties, and applications. *Carbohydrate Polymers.* 181:851-861.

Harrington, J.F. 1972. Seed storage and longevity. In T.T. Kozlowski, (Ed) *Seed Biology,* Vol 3 pp. 145-245 Academic Press, Inc., New York.

Hay, F.R., P. Thavong, P. Taridno, and S. Timple. 2012. Evaluation of zeolite 'Drying Beads®' for drying rice seeds to low moisture content prior to long-term storage. *Seed Science and Technology* 40:374-395.

Herter, U. and Burris, J. S. 1989. Effect of drying rate and temperature on drying injury of com seed. Can. J. Plant Sci. 69: 763-'I'74.

Hill, M.J. and C.R. Johnstone. 1985. Heat damage and drying effects on seed quality. *New Zealand Grassland Association: Research and Practice Series.* 2:53-57.

ISTA (International Seed Testing Association). 2005. International Rules for Seed Testing. Edition 2005. International Seed Testing Association, Bassersdorf, Switzerland.

Justice, O.L., L.N. Bass, and United States Science and Education Administration. 1978. Principles and Practices of Seed Storage. United States Department of Agriculture, Science and Education Administration. Agriculture Handbook No. 506.

Kew Royal Botanical Gardens. 2022. Small-scale seed drying methods. Technical Information Sheet 08.

Kiefer, C., T. Motis, and E. Toevs. 2020. Earthbag Seed Banks. *ECHO Technical Note* no. 96.

Kunze, O.R. 1977. Moisture adsorption influences on rice. *Journal of Food Process Engineering* 1(2):167-181.

Lawrence, B., A. J. Bicksler, and K. Duncan. 2017. Local Treatments and Vacuum Sealing as Novel Control Strategies for Stored Seed Pests in the Tropics. *Agronomy for Sustainable Development*. 37 (6): 1-11.

Malik, M. R. Sindhu, S.B. Dhull, C. Bou-Mitri, Y. Singh, S. Panwar, and B.S. Khatkar. 2023. Nutritional composition, functionality, and processing technologies for amaranth. *Journal of Food Processing and Preservation*. 2023:1-24.

Martineau-Côté, D. A. Achouri., S. Karboune, and L. L'Hocine. 2022. Faba bean: an untapped source of quality plant proteins and bioactives. *Nutrients*. 14:1541.

McCormack, J.H. 2004. Seed processing and storage: Principles and practices of seed harvesting, processing, and storage: an organic seed production manual for seed growers in the Mid-Atlantic and Southern U.S. Carolina Farm Stewardship Association.

McDonald, M. B. 2007. Seed moisture and the equilibrium seed moisture content curve. *Seed Technology*. 29(1):7-18.

McDonald, M.B. and L.O. Copeland. 1997. "Drying and Storage," in Chapter 7 of *Seed Production Principles and Practices*. Springer Science+Business Media Dodrecht.

Meitner, L.S. and V. Ricciardi. 2012. Seed fairs: Fostering local seed exchange to support regional biodiversity. *ECHO Asia Notes* no. 12.

Meitzner, L.S. and M.L. Price. 1996. *Amaranth to Zai Holes: Ideas for growing food under difficult conditions.*

Morgan, S. 2013. Parched Rice Desiccant. An ECHO video on heating brown rice over an open flame.

Motis, T.N. 2016. Seed Storage in the Tropics. *ECHO Best Practice Notes* no. 5. https://www.echocommunity.org/en/resources/6fa5029a-b130-4561-aa58-b2bf117de358

Motis, T.N. and D.R. Berkelaar. 2012. *Agricultural Options for Small-Scle Farmers: A Handbook for Those Who Serve Them.*

Mrema, G.C. 2011. *Rural structures in the tropics: design and development*. Rome: Food and Agriculture Organization of the United Nations, pp. 363-364.

Organic Seed Alliance. (2010). A seed saving guide for gardeners and farmers. Townsend, WA: OSA. www.seedalliance.org

Oyekale, K.O. C.C. Nwangburuka, O.A. Denton, J.A. Adeyeye, S.E. Ayeni, and O.K. Raheem. 2014. Predicting the longevity of sesame seeds under short-term containerized storage with charcoal desiccant. *American Journal of Experimental Agriculture* 4(1):1-11.

Pellegrini, P. A. and Galo E. Balatti. 2016. Noah's Arks in the XXI Century. A Typology of Seed Banks. Biodiversity & Conservation. 25: 2753-2769.

Potts, H.C. 1972. A closer look at seeds. *Seed Technology Papers*. 3.

Price, Z. 2016. Constructing an Improved Cold Room for Seed Storage. *ECHO Asia Notes* no. 27. https://www.echocommunity.org/en/resources/3171bfa2-b742-40d2-9ae1-5dc4da827628

Rao, N. K., J. Hanson, M. E. Dulloo, K. Ghosh, D. Nowell, and M. Larinde. 2006. *Handooks for Genebanks No. 8: Manual of Seed Handling in Genebanks*. Rome: Biodiversity International.

RBG Kew. 2016. The State of the World's Plants Report – 2016. Royal Botanic Gardens, Kew

Reader, S. and T. Motis. 2017. Are My Seeds Dry Enough? *ECHO Development Notes* no. 136.

Steil, A. and R. Jauron. 2023. Germination Requirements for Annuals and Vegetables. I*owa State University Extension and Outreach: Horticulture and Home Pest News*.

Sutcliffe, V. and J. Adams. 2014. Low-cost monitors of seed moisture status. Technical Information Sheet, 7. Royal Botanic Gardens, Kew.

Trail, P., T. Motis, S. Swartz, and A. Bicksler. 2022. Low-Cost Seed Storage Technologies for Development Impact of Small-Scale Seed Saving Entities in Tropical Climates. *Experimental Agriculture*. 57:324-337.

Trail, P. J. and B. Thansrithong. 2021. Storing *Lablab purpureus* (L.) seed under vacuum using a modified bicycle pump. *Proceedings of the V Asia Symposium on Quality Management in Postharvest Systems.* 1336: 243-248.

Trail, P., Y. Danmalidoi, M. Pler, A. Bicksler, and B. Thansrithong. 2019. Low-Cost Natural Building Options for Storing Seed in Tropical Southeast Asia. *ECHO Asia Notes*. no. 38.

Trusiak, M. B.P. Plitta-Michalak, and M. Michalak. 2023. Choosing the right path for the successful storage of seeds. *Plants* (Basel). 12(1):72.

Tshin, R. 2013. How to Facilitate Seed Exchanges During Country Meetings or as a Single-Day Event. *ECHO Asia Notes* no. 16.

VanNocker, A. and V. Reed 2012. Annual Leafy Greens Variety Trial. *ECHO Development Notes* no. 117

Vernooy, R., Sthapit, B. and Bessette, G. 2020. Community seed banks: concept and practice. Facilitator handbook (updated version). Rome (Italy): Bioversity International 150 p. ISBN: 978-92-9255-068-4 https://hdl.handle.net/10568/81286

Warneke, L. 2020. Silica gel FAQs. https://laboratoryresource.com.au/?navaction=getitem&id=155

Wilnon, R.L. 1987. Chapter 7. Seed Storage. In: *A guide to forest seed handling*. Food and Agriculture Organization.

World Benchmarking Alliance. 2019. Access to Seed Index 2019: A Synthesis Report.

Made in the USA
Monee, IL
02 November 2026

67378089R00083